# MACHINE LEARNING

## Concepts, Tools and Data Visualization

# MACHINE LEARNING

## Concepts, Tools and Data Visualization

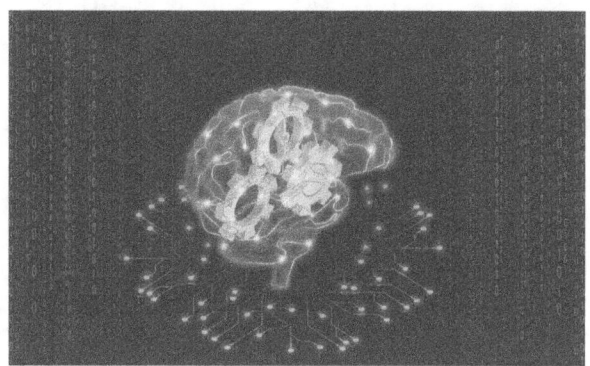

## Minsoo Kang
Eulji University, South Korea

## Eunsoo Choi
All4Land Inc., South Korea

**World Scientific**

NEW JERSEY · LONDON · SINGAPORE · BEIJING · SHANGHAI · HONG KONG · TAIPEI · CHENNAI · TOKYO

*Published by*

World Scientific Publishing Co. Pte. Ltd.

5 Toh Tuck Link, Singapore 596224

*USA office:* 27 Warren Street, Suite 401-402, Hackensack, NJ 07601

*UK office:* 57 Shelton Street, Covent Garden, London WC2H 9HE

**British Library Cataloguing-in-Publication Data**
A catalogue record for this book is available from the British Library.

**MACHINE LEARNING**
**Concepts, Tools and Data Visualization**

ISBN 978-981-122-814-8 (hardcover)
ISBN 978-981-122-936-7 (paperback)
ISBN 978-981-122-815-5 (ebook for institutions)
ISBN 978-981-122-816-2 (ebook for individuals)

For any available supplementary material, please visit
https://www.worldscientific.com/worldscibooks/10.1142/12037#t=suppl

Typeset by Stallion Press
Email: enquiries@stallionpress.com

# About the Author

**Prof. Min-soo Kang**

Professor Min-soo Kang, who is in the Department of Medical IT at Eulji University, serves as the Director of the Computing Center at the Eulji University Medical Center and the Editor-in-Chief of the Korean Artificial Intelligence Association. He is working on machine learning in the medical field with the Division of Cardiology, Division of Rheumatology, and Division of Laboratory Medicine at Eulji University Hospital. Professor Kang received a Ph.D. from the Department of Control and Instrumentation Engineering at Kwangwoon University and is currently studying IoT and artificial intelligence. He is an adjunct professor at Hanyang University and a professor of Information and Communication Engineering at Hanyang Cyber University. He also served as the RFID/USN Program Director in the Ministry of Knowledge Economy, which conducts the planning and evaluation of national R&D projects. His book publications are Getting Started with Machine Learning using Microsoft Azure Machine Learning, RFID Basics (a textbook approved by the Superintendent of Education in Seoul), Digital Circuit Design using VHDL, and RFID GL Qualification Test (Nationally Authorized) and so on.

**Assistant Research Engineer Eun-Soo Choi**

Assistant Research Engineer Eun-Soo Choi, who received a B.S and an M.S degree from the Eulji University Department of Medical IT Marketing, wrote a thesis on data preprocessing and hybrid machine learning. He finished the Microsoft Deep Learning (CNTK) course and Azure Machine Learning Studio course. He has gone through MIIC (Medical Intelligence Information R&D

Center) and is currently serving as a military service exempted technical research personnel at the Smart Geospatial Research Center. Now, he is working on a deep learning-based Point Cloud modeling project. Also, he is working on GIS and BIM-based machine learning applications. His book publication is Getting Started with Machine Learning using Microsoft Azure Machine Learning.

# Preface

Artificial intelligence, which is the primary driver of the fourth Industrial Revolution, will develop our society into an intelligent society where everything from industry to education to medicine is interconnected. Artificial intelligence is increasing human potential by creating machines which simulate human intelligence. Some examples of artificial intelligence successes include Deep Blue's chess victory in 1997 and AlphaGo's Go victory in 2015, the test drive success of autonomous vehicles, the development of artificial intelligence speakers with voice recognition capabilities which can respond to voice commands, and the implementation of artificial intelligence diagnosis support to help doctors. As many exciting developments are taking place in the field of artificial intelligence, the number of researchers in the field is increasing; still, terms such as 'machine learning' and 'deep learning', along with ambiguous definitions, are making it difficult for researchers to enter the field. The author, too, remembers how difficult it was to enter the field; he studied control theory in robotics. One thing for sure is that even non-artificial-intelligence majors want to use artificial intelligence.

For those who start on artificial intelligence without any particular application in mind, they will think that academic approaches are necessary. These may range from the history of artificial intelligence to the definition of artificial intelligence terms to the logic, mathematics, and algorithms of artificial intelligence. However, those who start on artificial intelligence with an interest in its applications will probably feel that that practical matters such as coding and algorithms, which can be applied to their fields, are of greater interest. I wrote a book which I hope will appeal to both these groups of people who have no expertise in artificial intelligence.

This book, firstly, was written for those who do not know mathematics and the theory of control, secondly, for those who tried the open-source approach but had difficulties.

People who don't know mathematics or control theory will be told what AI is and how data can be used in the application of AI. For best results, they should follow the instructions in the book. Those who use open-source approaches can read the text carefully and decide whether to choose the open-source approach or not. There are pros and cons to choosing an open-source approach. Because it is open-source, the approach will cost nothing, but there will be limitations in applying the required libraries or algorithms. The time spent finding alternative ways to perform tasks could be better invested in actual research. The difference between closed-source and open-source approaches is similar to the difference between Windows/iOS users and Linux users. Windows/iOS users pay for their OS but have a greater variety of programs and access to support. Linux users, on the other hand, do not pay for their OS, but their OS is less convenient and more difficult to use. I hope you understand my analogy explaining the difference between open-source and closed-source approaches. For this book, I have used Microsoft's Azure Machine Learning program.

This book consists of three parts. Part 1 briefly introduces the history of artificial intelligence, machine learning, and deep learning, as well as algorithms. It also explains how the data will be utilized i.e. in data science approaches such as data mining and big data processing. Data mining and big data processing are ways of discovering patterns in large data sets involving methods at the intersection of machine learning, statistics, and database systems.

Part 2 introduces Azure Machine Learning, how to use it, and some examples as needed. Azure Machine Learning makes it easy to use algorithms used in supervised and unsupervised learning by selecting menus. So, Azure Machine Learning is explained and each step is illustrated so that we can study the example program from the beginning to the end. We also described examples of using actual public and medical data. If this book is read in its entirety, even beginners who have just started machine learning can use Azure Machine Learning programs.

Part 3 introduced visualization and Power BI, which is a visualization tool. Data visualization is an essential technique in all areas of data handling. Visualization will help you lay the foundation of data science. Power BI was developed by Microsoft, and so can be integrated with Microsoft Azure Machine Learning. Also, Power BI's R Script feature enables complex computations for high-performance visualization.

If you want to apply artificial intelligence rather than follow the theoretical approach, you will have to learn a program, not necessarily Azure Machine

Learning Studio. Local and overseas laboratories have developed many artificial intelligence programs, but the Azure ML Studio program, chosen by the author, offers a wide variety of functions, which will be required in this book. However, you need to know that licensing costs are incurred. Here's a tip for beginners who are starting Azure Machine Learning: we recommend that you take advantage of the 8-hour trial version. Readers should expect to be able to understand AI and visualize the results of using AI on their data by the end of the book.

Before we begin, I would like to express my appreciation for WSP officials and especially Sunny Chang for their help in publishing this book. I would also like to thank Professor Fergus Lee Dunne, Song Seo-won, and Kwak Young Sang of the Medical Intelligence Information Center.

# Book Reviews

**Cisco Systems Korea, Vice President, Sin Ui Cho**
The history of machine learning is short but well-organized and explains Microsoft's approach to the Microsoft Azure Machine Learning Studio program. Even people who don't know a programming language can easily program if they are interested in machine learning. In Part 2, an example program is written step by step with Microsoft Azure Machine Learning Studio. This example can be followed by the reader in order to get results. In addition, it is expected that the use of real data will be helpful to researchers in data science and analytics.

**Microsoft Korea IoT Solution Architect, Executive Managing Director, Geon Bok Lee**
While many books already explain AI, it was hard to find a book that is appropriate for developers in the age of AI's democratization. This book provides a systematic approach to making AI easy to use for developers who do not have a deep knowledge of AI statistics. Reading this book is the best way to gain the vivid experience and know-how of a professional who has worked with Microsoft for years.

**LGU+, Senior Vice President/Enterprise Sales Group2, Sunglyul Park,**
"Starting Machine Learning using Microsoft Azure Machine Learning Studio" begins with a history of artificial intelligence and proceeds to summarize the current state of knowledge in machine learning. It teaches the beginner in artificial intelligence how to write artificial intelligence programs and organize data, even if he or she does not possess the mathematical knowledge needed. The part about the visualization of data teaches the reader how to present their data analysis in a way that is easy to understand.

**Microsoft Korea GBB AA/AI (Advanced Analytics/
Artificial Intelligence), General Manager, Seok Jin Han**

Machine learning is rapidly spreading across various industries and is expected to become a necessary factor for the competitiveness of companies and individuals in the future. This book will help you learn the basic concepts of machine learning quickly through Microsoft Azure Machine Learning Studio and R. If the practical information in this book is supplemented with additional learning materials on the concepts/technology/algorithms used in machine learning, the reader can maximize his or her learning opportunities.

**Yonsei University, Assistant Professor, Ki Bong Yoo**

This book was written to help people who do not have math or programming skills to pick up machine learning. In particular, there is a part about visualization that makes it easier to understand the results of data analysis.

**KIBA (Korea Institute of Business Analysis & Development),
CEO, Min Sun Kim**

Artificial intelligence is being applied in all fields of the industry, and especially in the security field. The artificial intelligence described in this book is well organized so that people who don't know programming or math can understand it. By the end of this book, even those who have not studied artificial intelligence in the field of engineering will be able to utilize artificial intelligence.

**PNP Secure, CEO, chun oh Park**

This book has the basic concept of artificial intelligence, and even people without an engineering background can run artificial intelligence programs. An artificial intelligence program is possible, statistics can be analyzed with the R program, and visualization is also made so that the entire cycle from a collection of artificial intelligence data to visualization can be easily progressed.

# Contents

# Part I
# Artificial Intelligence

# Chapter 1

# Summary of Artificial Intelligence

## 1.1 Definition of Artificial Intelligence

Artificial intelligence has several definitions, including "machine-made intelligence" and "an artificial embodiment of some or all of the intellectual abilities possessed by humans."

The various definitions of artificial intelligence can be summarized as follows: "machines equipped with human intelligence capable of understanding human judgment, behavior, and cognition". Intelligence is said to be "the ability to apply prior knowledge and experience to achieve challenging new tasks." This, ultimately, can be said to refer to human intellectual ability. This ability can be used to respond flexibly to a variety of situations and problems and is also related to learning ability.

Learning ability is the ability to acquire skills or information that is difficult for others to learn, or to learn the same content faster or more extensively. While intelligence itself is different from what an individual has learned, academic achievement, intellectual thinking and behavior depend on prior knowledge; so, intelligence can be changed through experience and learning.

Therefore, intelligence generally starts with, and is defined as, "the ability to solve problems." We need a high level of intelligence to solve various problems. For example, when you play chess, you will use information acquired from a variety of texts, numerous chess games, and the Internet to make predictions and then act accordingly. These behaviors are highly demanding of human intelligence and thought emerges as a result of this intelligence-based learning. Scientific explanations of "thought", broadly define it as "consciousness", comprehensively identifying it as "experience" of an object. Experience is also knowledge of a process, about which a person, who has had considerable experience in a field, may learn in order to gain a reputation as an expert. Thus, intelligence is the recognition, analysis, and understanding of the thoughts and experiences that human beings may have, while that which is produced artificially is termed artificial intelligence. To create a machine that thinks like a human, it is necessary to study human thoughts and behaviors, including listening, speaking, seeing, and acting. It may seem easy, but for a computer to have the ability to listen, speak or act based on what it has heard or seen is a very difficult task, and so it would be faster to emulate what humans think.

So, as a scientific approach, human-like thoughts and actions are required. To put this into practice, much like a child learns experientially by thinking and acting on their own, there is a need for the thorough study and investigation of processes of repeated learning and experimentation. Prof. Kathleen McKeown distinguishes four main categories of human thought and behavior.

- **Systems that think like humans**

This is a cognitive modeling approach used with a system that is capable of human-like thinking and decision-making as defined by Hodgeland in 1985 and Bellman in 1978. Haugeland (1985) described this approach as "the exciting new effort to make computers think…machines with minds, in the full and literal sense", and Bellman (1978) described it as "the automation of activities that we associate with human thinking, activities such as decision-making, problem solving, learning". The aim of this approach is to create a computer model based on artificial intelligence which can mimic human thought through actual experiments.

- **Systems that think rationally**

This is a law-of-thought approach to thinking used with a system with mental abilities such as perception, reasoning, and behavior which are based on a calculation model, as defined by Charniak and McDermott in 1985 and Winston in 1992. There are two main obstacles to this approach. First, it is not easy to take informal knowledge and state it in the formal terms required by logical notation, particularly when one is less than 100% certain of the knowledge being handled in this way. Second, there is a big difference between being able to solve a problem "in principle"' and doing so in practice. Charniak and McDermott (1985) have studied the simulation of mental faculties using computational models, and Winston (1992) has made a study of the computations that make it possible to perceive, reason, and act. The key to this approach is to format non-formatted items such that they can be handled by the machine's logical system, which can then draw its own conclusions as to what should be done next. In this way, the human thought process can be programmed into a computer.

- **Systems that act like humans**

This is a Turing test approach, defined by Kurzweil in 1990 and Rich and Knight in 1991, featuring a system that allows the machine to mimic any action that requires human intelligence. Alan Turing's proposed test in 1950 is a method for determining whether a machine has intelligence based on how similarly to a human the machine behaves when communicating with an actual human. That is, to pass the Turing test, the machine must achieve the same level of performance as a human in a conversation, such that the human testing it believes it is a human. Kurzweil (1990) calls this "the art of creating machines that perform functions that require intelligence when performed by people", while Rich and Knight (1991) calls it "the study of how to make computers do things at which, now, people are better".

- **Systems that act rationally**

This is a rational agent approach featuring an agent system that acts intelligently via a calculation model, as defined by Schalkoff in 1990 and Luger and Stubblefield in 1993. An agent is something that perceives and acts; as such, computer agents are not just "programs," but are unique in that they operate under autonomous control, perceive the environment, survive for a long time, adapt to change, and mimic the goals of other agents. Schalkoff (1990), quoted in Pool *et al.* (1998) places this approach within "a field of study that seeks to explain and emulate intelligent behavior in terms of computational processes" while Luger and Stubblefield (1993), quoted in Nilsson (1998) places it within "the branch of computer science that is concerned with the automation of intelligent behavior". A rational agent is one that acts to get the best results, or, in cases of uncertainty, the best-expected results. Artificial Intelligence, as a form of rational agent design, therefore, has two advantages. First, it is more general than the "laws of thought" approach, because correct inference is only a useful mechanism for achieving rationality, and not a necessary one. Second, it is more amenable to scientific development than approaches based on human behavior or human thought, because the standard of rationality is clearly defined and completely general.

In conclusion, artificial intelligence is a technological or scientific field that studies the methodology or feasibility of producing intelligence by implementing machine-made human cognition, reasoning, and learning into computers or systems.

## 1.2 History of Artificial Intelligence

### 1.2.1 *The Beginning of Artificial Intelligence*

The birth of artificial intelligence began in 1943 when Warren McCulloch and Walter Pitts began studying artificial intelligence. Their study of AI was prompted by three sources: knowledge of physiology and neuronal function in the brain; formal analysis of propositional logic according to Russell and Whitehead; and Turing's Theory of Computation.

Their proposed model of artificial intelligence was based on a connection model in which neurons[1] process information with synapses. A neuron is marked on or off when stimulated by another connected neuron. This model represents the working relationship between neurons, which theoretically proves that by connecting the neurons in an artificial neural network in the form of a net, one can mimic how the simpler functions of the human brain operate. In terms of artificial intelligence, the "neurons" used in neural networks were the first neuron models. Such a "neuron" was conceptually defined as "a proposition that provides sufficient stimulation".

McCulloch and Pitts also argued that the concept of learning can be used to describe neuron-connected networks; in 1949, Hebb named this concept "Hebb's learning rule."[2] In 1950, Alan Turing developed the Turing test while the feasibility of exhibiting machine intelligence was being discussed, and in 1951, Christopher Strachey and Dietrich Prince launched their chess program on the Ferranti Mark 1 machine at Manchester University. Subsequently, in 1959, Arthur Samuel stated that, "a field of research that develops algorithms that allow machines to learn from data and execute actions that are not explicitly specified by code" was enough technological advancement to define machine learning and challenge researchers in the field. Figure 1-1 illustrates the history of artificial intelligence.

The 1956 Dartmouth Artificial Intelligence conference marked the beginning of the field of AI and provided succeeding generations of scientists with their first sense of the potential information technology had to be of benefit to human beings in a profound way. Figure 1-2 shows the participants of Dartmouth's conference. In the summer of 1955, John McCarthy, then associate professor of mathematics at Dartmouth College, decided to organize a group to clarify and develop ideas about thinking machines. In the early 1950s, there were various names for the field of "thinking machines": cybernetics, automata theory, and complex information processing. However, instead of focusing on automata theory, John McCarthy chose a new name, artificial intelligence. The project was formally proposed by McCarthy, Marvin Minsky, Nathaniel Rochester and Claude Shannon on September 2, 1955. The proposal is credited with introducing the term 'artificial intelligence'.

---

[1] Neurons (also called neurons or nerve cells) are the fundamental units of the brain and nervous system, the cells responsible for receiving sensory input from the external world, for sending motor commands to our muscles, and for transforming and relating the electrical signals at every step in between.

[2] Hebb's learning rule is a postulate proposed by Donald Hebb in 1949. It is a learning rule that describes how the neuronal activities influence the connection between neurons, i.e... the synaptic plasticity. When an axon of cell A is near enough to excite a cell B and repeatedly or persistently takes part in firing it, some growth process or metabolic change takes place in one or both cells such that A's efficiency, as one of the cells firing B, is increased.

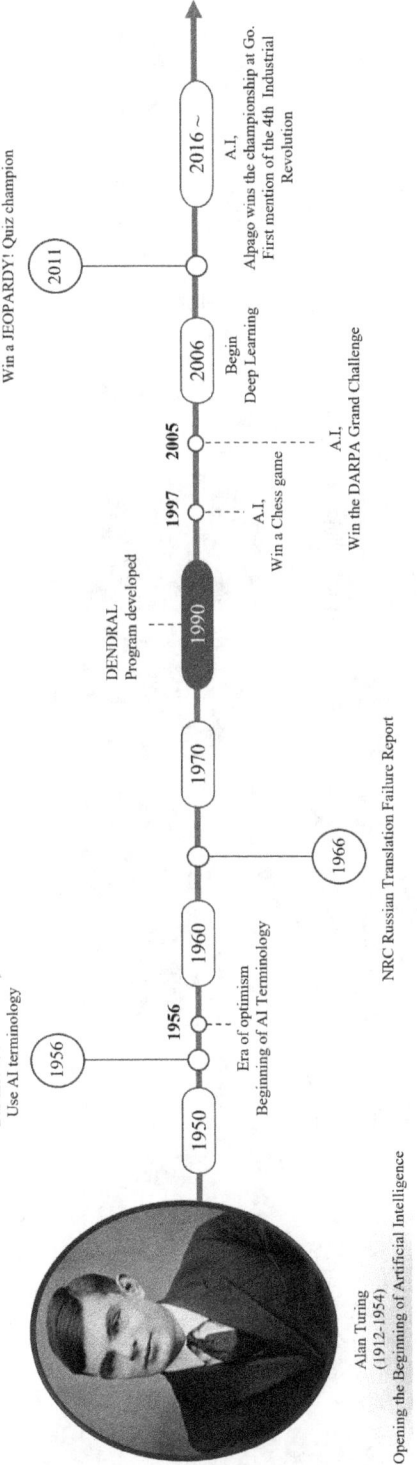

**Figure 1-1** History of artificial intelligence

**(a) Dartmouth Conference Participants**

Jonh McCarthy    Marvin Minsky    Claude Shannon    Ray Solomonoff    Alan Newell

Herbert Simon    Arthur Samuel    Oliver Selfridge    Trenchard More    Nathaniel Rochester

**(b) AI founder of Dartmouth Conference**

**Figure 1-2**    Dartmouth's conference participants

**The Proposal states**

We propose that a 2-month, 10-man study of artificial intelligence be carried out during the summer of 1956 at Dartmouth College in **Hanover, New Hampshire**. The study is to proceed on the basis of the conjecture that every aspect of learning or any other feature of intelligence can in principle be so precisely described that a machine can be made to simulate it. An attempt will be made to find how to make machines use language, form abstractions and concepts, solve kinds of problems now reserved for humans, and improve themselves. We think that a significant advance can be made in one or more of these problems if a carefully selected group of scientists work on it together for a summer.

The proposal discusses computers, natural language processing, neural networks, theories of computation, abstraction and creativity. While AI did not yield significant results in the immediate wake of Dartmouth's workshop, in time the workshop not only provided a direct opportunity for the study of various areas of artificial intelligence, it would also set the stage for the next phase of research at MIT, Carnegie Mellon University, Stanford University, and IBM in the United States.

## 1.2.2 *Early Artificial Intelligence*

The next stage of computer and programming tool development featured the emergence of primitive but elementary computing power. At this point, Nowell and Simon had been very successful in developing a system called the General Problem Solver (GPS) that could change many of the factors in calculations, thereby reducing the difference between desired and expected results. The GPS is a program that models the process of solving human problems, unlike the LT (Logic Theorist). It is an inference program that uses a combination of searching, goal-oriented behavior, and rule-application.

GPS became the first program to achieve the first goal of artificial intelligence — a system that can think like a human. Beginning at Carnegie Mellon University, the cognitive science approach and convergence research, which suggests that human cognitive skills will eventually become identical with those of computers, is still the basis of cognitive science today. These approaches are also studied in the fields of biology, linguistics, psychology and computer science.

McCarthy, who announced the beginning of AI, made three critical achievements when he moved from Dartmouth to MIT in 1958. First was the development of the artificial intelligence programming language LISP, which became the second oldest programming language. Second was the introduction of Time Sharing, which began to solve problems relating to the expensive nature of computer usage. At the time, the fee for using a computer was high; however, Time Sharing meant that multiple users were able to take turns using one computer concurrently for short time periods. Third, in 1958, McCarthy published his first book, "Programs with Commonsense," which led to the development of the first complete AI program called "Advice Taker."

This program included all the essential principles of knowledge representation and reasoning, and knowledge was used in the search process to solve problems.

### 1.2.3 *The Stagnation of Artificial Intelligence*

From the beginning, artificial intelligence researchers predicted the success of their field, with Herbert Simon even announcing in 1957 that "the world now has machines to think, learn, and create". Simon's overconfidence was largely based on the ability of artificial intelligence systems to perform successfully in simple situations. In almost all cases, however, when these early systems had to choose from a broader range of problems, it became more difficult for them to perform successfully, and they revealed their limitations. The reasons for this are as follows.

First, the early artificial intelligence programs only succeeded in simple grammatical manipulation because they had a limited knowledge of the subject. The National Research Council in the United States financed the translation of the Russian language, but it was difficult to resolve the ambiguity of the language and to understand sentence content. Eventually, in 1966, the Advisory Committee issued a report saying that "there is no machine translation of the general scientific text and no immediate outlook." The government's financial support for academic translation projects was then canceled.

Second, the problems that artificial intelligence was trying to solve were complicated.

The early artificial intelligence programs were able to resolve the problem step by step by expressing only the basic facts about it until a solution was found. A simple program could solve the problem by finding a solution, but a complex AI program was impossible.

As such, at the time, it was hoped that a better computer machine or improvements in memory would become available. However, as the Computational Complexity Theory developed over time, it was found that solving one problem using just one program did not linearly scale in time and space, depending on the difficulty of the problem.

Third, the basic structure used to generate intelligent behavior was limited.

For example, according to Minsky and Papert's (1969) *Perceptrons: An Introduction to Computational Geometry*, it was impossible for neural networks to learn certain facts, which proved to be a problem. Of course, over the years, the Back-Propagation learning algorithm for multi-layer neural networks would end up being introduced to solve this problem.

## 1.2.4 *The Reactivation of Artificial Intelligence (1969–1990)*

There were many reasons why AI was challenging to develop; however, the general-purpose search method provided a solution to these problems. The reason for these problems centered around the paucity of knowledge of the parts of a problem to be solved, while performance was inevitably reduced when solving complex parts of problems. This approaches are called "weak methods," which, although common, made it difficult to solve for larger or more difficult problems.

The solution was to use more powerful domain-specific knowledge to make it easier to deal with cases that typically occur, and to allow for more significant inference. In this respect, in cases where someone is trying to solve a difficult problem, there is an eventual realization that they can only solve it effectively by applying knowledge gleaned from someone who already knows the answer: in other words, by using expert knowledge.

Buchanan's "DENDRAL" program, announced around this time, is a prime example.

The DENDRAL project was one of the earliest expert systems, and its primary aims were researching the method of scientific reasoning and the formalization of scientific knowledge by working in the specific field of organic chemistry. DENDRAL proposed the most likely plausible candidate structures for new and unknown chemicals, and in some classes of organic materials, their performance measured up with that of human experts. The importance of DENDRAL lies in the notion that it was probably the first-ever successful knowledge-intensive system.

## 1.2.5 *The Augustan Era (Platinum Age) of Artificial Intelligence (1980–present)*

The first commercially successful expert system was R1, used by Digital Equipment Corporation. The program helped with the process of ordering computer parts and peripherals when introducing a new computer system. By 1986, about $40 million had been saved each year. By 1988, DEC's AI group was using an additional 40 expert systems.

Recently, neural networks began to flourish and have been applied to various fields.

While neural networks had been abandoned by the end of the 1970s, work continued in other fields. Physicists such as John Hopfield treated a set of nodes as a set of atoms, analyzing storage and optimizing networks using statistical mechanics techniques.

These neural networks were also applied to communications networks to predict network traffic in advance in order to minimize communication load and operate efficiently. Also, a great deal of research has been conducted on intelligent agents with intelligence agent theory. In particular, smart mobile systems in information retrieval systems are widely used and progress is being made and applied to numerous industrial operations around the world.

# 1.3 Classification of Artificial Intelligence

From a philosophical perspective, AI can be divided into strong AI and weak AI. Strong artificial intelligence is defined as computer-based artificial intelligence that can think and solve problems, while weak artificial intelligence is defined as computer-based artificial intelligence that cannot do either of these tasks.

## 1.3.1 *Strong Artificial Intelligence*

The qualities of perception and self-awareness can then be said to typify strong forms of artificial intelligence. There is, therefore, a difference between this form and weak artificial intelligence, in that the former can actually think or solve problems. It is a form of artificial intelligence in which computer programs can act or think in a human-like manner.

In other words, strong artificial intelligence is not simply a computer, but a computer that can think with human intelligence. Much like the machines in the movie "Terminator" that caused a war with humans, not only can strong artificial intelligence work on its own without being commanded, it can reject orders it deems to be irrational. Thus, there are two forms of AI; human-like AI; which involves computer programs behaving like humans and thinking human-like thoughts; and non-human AI, which involves computer programs developing perceptions and inferences that are different from humans.

- **Artificial General Intelligence (AGI)**

In the term Artificial General Intelligence, "General" should be understood as meaning "general" rather than 'normal.' In other words, unlike artificial intelligence that can be applied only under specific conditions, Artificial General Intelligence refers to artificial intelligence that can be used in all situations.

Artificial General Intelligence, therefore, unlike weak artificial intelligence, relates to a machine's ability to see and learn from others and to perform tasks that they have not done before. It is a form in which, by seeing and hearing, self-learning is possible. General Artificial Intelligence is the form of intelligence aimed for by AI researchers these days who have the goal of "making machines behave like humans".

- **Artificial Consciousness (AC)**

Artificial Consciousness is one form of artificial intelligence and is also called machine consciousness or synthetic consciousness. It can be thought of as the next form of artificial general intelligence. If artificial intelligence analyzes and understands ordinary objects, Artificial Consciousness goes beyond this to imitate or even possess emotions, an ego, or creativity. It is also a form of artificial intelligence which can control itself, making judgements about its environment and actively making use of certain objects by thinking about what is necessary to perform a task, even without being given specific orders.

Therefore, artificial consciousness presupposes intelligence that is similar to or surpasses that of humans, and, unlike artificial intelligence, which is merely the result of programming a computer to execute a given task, artificial consciousness requires more than mere instruments. In fact, artificial consciousness requires an intelligence of the same order as humans. It may be assumed that artificial intelligence will reach the level of artificial consciousness at a future point in its development, but at this point, it should be considered as a hypothesis.

In the end, the question of whether it is possible to implement a consciousness that thinks like humans mechanically, and whether a human-like machine can theoretically be built, is a fundamental debate in artificial intelligence circles. It is also a debate with significant ethical implications.

## 1.3.2 *Weak Artificial Intelligence*

Weak artificial intelligence is defined as artificial intelligence that focuses on solving specific problems with the use of computers that cannot think or solve a problem in actuality. Although they are not intelligent in actuality, or in possession of what might be termed intelligence, the development of programs that mimic human intelligence through the use of a set of predefined rules produces behavior that resembles intelligent behavior. While research in the field of strong artificial intelligence is often seen as the path most favored by current artificial intelligence researchers, it has its drawbacks; however, a fair amount of progress has been made in the field of weak artificial intelligence which focuses on a specific aim.

Google's AWS, DeepMind's AlphaGo and IBM's Watson are all examples of rule-based artificial intelligence, and thus, examples of weak artificial intelligence.

Optimized artificial intelligence is a type of artificial intelligence technology that solves certain types of problems or completes a task. As already mentioned, weak AI requires the input of algorithms, basic data, and rules. Today's weak artificial intelligence is widely used in medicine, management, education, and services; unlike strong artificial intelligence that can think for itself in a human-like fashion, it is restricted to thinking and operating within narrow areas of human cognitive ability.

# 1.4  Practice Questions

Q1. What is artificial intelligence?

Q2. Prof. Kathleen McKeown divided human thoughts and behaviors into four systems. Describe the four systems.

Q3. Describe the proposal for the Dartmouth workshop.

Q4. John McCarthy, who announced the beginning of artificial intelligence, achieved three important things when he moved from Dartmouth to MIT in 1958. Describe McCarthy's three achievements.

Q5. From a philosophical perspective, AI can be divided into strong AI and weak AI. Describe strong AI and weak AI.

# Chapter 2

# Machine Learning

## 2.1 Definition of Machine Learning

Machine Learning (ML) is the scientific study of algorithms and statistical models that computer systems use to perform a specific task without using explicit instructions, relying instead on patterns and inference. It is seen as a subset of artificial intelligence. It also focuses on representation and generalization. Representation is the evaluation of data, and generalization is the processing of data that is not yet known. The term Machine Learning was first used by Arthur Samuel, an IBM researcher in the field of artificial intelligence, in his paper "Studies in Machine Learning Using the Game of Checkers."

The machine is a computer that can be programmed and used as a server. The study of machine learning is characterized by three approaches. The first is the neural model paradigm. Neural models, having started out in the form of the perceptron, are now moving towards deep learning models. Second is the study of the symbolic paradigm. This paradigm uses logic or graph structures instead of numerical or statistical theories and was the core approach of artificial intelligence from the mid-1970s up until the late 1980s. Third is the intensive paradigm of modern knowledge. This paradigm, which began in the mid-1970s, started with the theory that knowledge that has already been learned should be reused, rather than starting the process of learning from scratch, as neural models do.

In the 1990s, practical machine learning research became more mainstream than real-world approaches that focused on computer learning methodologies. In the 1990s, the paradigm of machine learning was largely computer based. Data mining, which analyzes data from a statistical perspective, shares a lot of theoretical elements with machine learning; the ease of securing digital data due to the rapid spread of high-performance computers and the spread of the Internet also had a lot of influence on this movement.

Regarding machine learning, Arthur Samuel said it was a research area designed to give computers the ability to learn without explicitly coding programs. Tom Mitchell stated, "A computer program is said to learn from experience E with respect to some class of tasks T and performance measure P, if its performance at tasks in T, as measured by P, improves with experience E." For example, when a computer is trained to recognize letters, the machine classifies the newly typed letters (T) and experiences learning from pre-processed data sets (E). To do (P) is to say that the computer "learned." Ultimately, it is a way to improve the performance of a task through experience.

### 2.1.1 *Machine Learning and Data Mining*

Machine learning is often mixed with data mining[1], probably because it uses the same methods, such as classification and clustering. In other words, solving problems using techniques, models, and algorithms such as classification, prediction, and clustering is called machine learning in the field of computer science, and data mining in the field of statistics. The combination of machine learning and data mining began in the 1990s when computer scientists sought more efficient solutions for handling statistics when researching practical machine learning.

The main difference between machine learning and data mining is that data mining focuses on discovering attributes, which are characteristics of data, while machine learning focuses on prediction using attributes known through training data.

Figure 2-1 shows that machine learning and data mining use the same algorithms, but computer science focuses on prediction, while machine learning focuses on attribute discovery.

## 2.2  Classification of Machine Learning

Machine learning distinguishes between supervised learning and unsupervised learning, depending on whether the learning data has a label. The most significant difference between supervised learning and unsupervised learning is that in supervised learning, a label is provided. A label is an attribute that defines the class an instance from the training data belongs to.

Figure 2-2 shows the classification of supervised and unsupervised learning.

---

[1] Data mining is the process of discovering patterns in large data sets involving methods at the intersection of machine learning, statistics, and database systems.

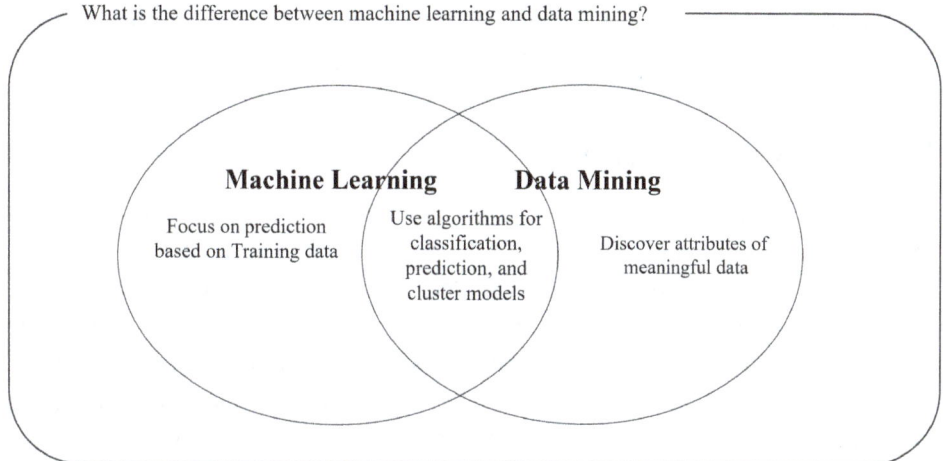

**Figure 2-1**  The difference between machine learning and data mining

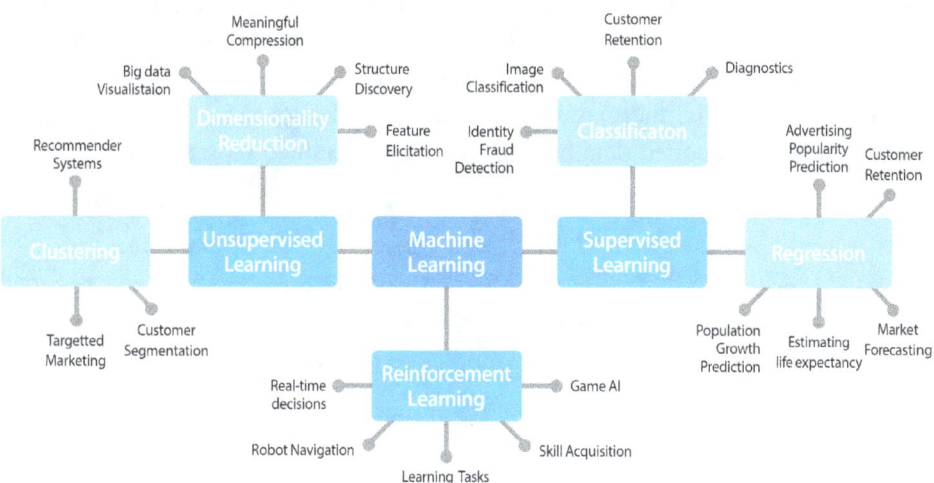

**Figure 2-2**  Classification of supervised and unsupervised learning

In order to teach a computer to recognize an object, one must show a real object or provide a picture of the object. The pictures or objects provided are training data, and the objects or objects in the pictures are pre-defined as 'cups,' 'desks,' 'bicycles', and 'cats.'.

Because labels are defined by a person's view of an object or picture, computers that learn from a person by reading the labeled pictures will classify objects in the same way as that person did.

On the other hand, if there is no label in the input data, it is called unsupervised learning because the computer has not been instructed by a human being.

Supervised learning has Classification, Regression, and Reinforcement models, while unsupervised learning has a Clustering model.

## 2.3 Supervised Learning

Supervised Learning is a learning model that takes a set of labeled training data and creates a discriminant that analyses the data using various algorithms and then uses the algorithms discovered on a new data set to produce results.

Supervised learning is divided into regression and classification. Since both classification and regression are supervised models, they have in common the fact that they learn from labeled input data. The difference between classification and regression is that classification results in a fixed value, whereas regression means that the resulting value can be any value within the range of the data set.

That is, the resultant value of the classification is one of the labels included in the training dataset, and the resultant value of the regression is any value calculated by a function expression (regression formula) determined by the training dataset. Figure 2-3 compares classification and regression.

More specifically, the classification model is designed to find the group to which the newly inputted data belongs after being trained on a data set in which

| | Classification | Regression |
|---|---|---|
| **Result** | Predict one of the labels in the training data (discrete) | Prediction of continuous values |
| **Example** | When the training data is A, B or C, the result is one of A, B, or C ex. Spam filtering on e-mail | The result can be any value ex. Stock price analysis |

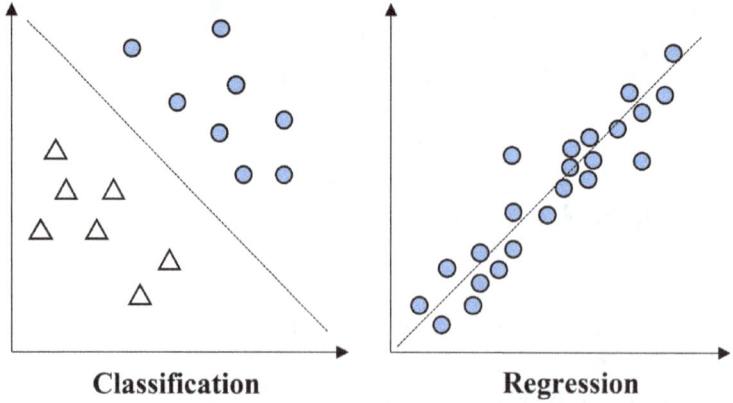

Classification            Regression

**Figure 2-3**    Comparison of classification and regression

each instance of data is assigned a group name (label). Therefore, the result value of the classification model must be one of the labels from the training data. For example, if you have a data set labeled "bicycle", "cat", "dog", "book", and "desk", the result of the classification model should be one of five labels i.e. bicycle, cat, dog, book, or desk. For reference, when a data set is classified into two groups, this is referred to as binary classification or binomial classification, and when a data set is classified into more than three groups, this is referred to as polynomial classification.

In statistical modeling, regression analysis is a set of statistical processes for estimating the relationships among variables. It includes many techniques for modeling and analyzing several variables, with the focus on the relationship between a dependent variable and one or more independent variables (or 'predictors'). The regression model aims to express the relationship between features and labels with the use of labeled training data. So, in a regression model that uses functional expressions, we look for specific patterns and predict what the dependent variables associated with particular independent variables should be according to those patterns. As such, it is much more reliable when predicting something because of its habit of returning to the mean average. Thus, regression analysis is also called a predictive model.

Regression analysis can also be applied to predict several categorical results, such as classification. The regression model in this case is called logistic regression. The logistic model is used to model the probability of a certain class or event occurring such as pass/fail, win/lose, or healthy/sick. This can be extended to model several classes of events such as determining whether an image contains a cat, dog, lion, etc.

## 2.3.1 *Classification*

In machine learning and statistics, classification is the problem of identifying to which of a set of categories (sub-populations) a new observation belongs, on the basis of a training set of data containing observations (or instances) whose category membership is known.

Classification refers to the process of recognizing, differentiating, and understanding a concept or subject, implying that a subject is in a category, and usually has a specific purpose. As such, classification is essential for all kinds of interactions concerning language, speculation, reasoning, decision making, and the environment. In machine learning, classification is a representative technique of data analysis, which classifies objects which have multiple attributes or variables into either a predetermined group or class.

Classification models are divided into various models according to the algorithms used. Among them, K-Nearest Neighbor, Support Vector Machine, and Decision Tree models are the most typical.

## (1) K-NN model

The K-NN (K-Nearest Neighbor) model used in classification is one of the most intuitive and simple of the machine learning models. The K-NN is a non-parametric method used for classification and regression. In both cases, the input consists of the k-closest training examples in the feature space. The output depends on whether K-NN is used for classification or regression: In K-NN classification, the output is a class membership. An object is classified by a plurality vote of its neighbors, with the object being assigned to the class most common among its k-nearest neighbors (k is a positive integer, typically small). If $k = 1$, then the object is simply assigned to the class of that single nearest neighbor. In K-NN regression, the output is the property value for the object. This value is the average of the values of k nearest neighbors. K-NN is a type of instance-based learning, or lazy learning, where the function is only approximated locally and all computation is deferred until classification. The reason for this is that the computer does not learn beforehand, but rather retains the data necessary for learning in its memory, and then starts to generalize when an instance occurs.

K-NN is an algorithm that classifies using labeled data.

For this purpose, the characteristics of the labeled learning data are quantified and then expressed in the coordinate space. The virtual circle extends around the new data point, searching for nearby data points to compare it to. When the nearest data point to the new data point is found, the group (label) to which the nearest data point belongs becomes the group associated with the new data point (when $k = 1$). In the same way, the virtual circle is expanded until more nearby data points are found. The new data point is then classified as belonging to the group that contains the most nearby data points. Figure 2-4 shows the classification process of the K-NN algorithm.

In K-NN, K looks for as many adjacent data as K to identify the group to which the new data belongs. But before selecting K, it is necessary to select the distance scale. Therefore, the concept of proximity should be based on a standard, and this is defined as Euclidean distance. When defining Euclidean distance, the unit must be considered. The units associated with coordinates must be distances, but other units can also represent money and weight, for instance. In addition, the distance in coordinates may change according to the unit. Therefore, the unit chosen with regard to the data is critical when defining the K-NN algorithm.

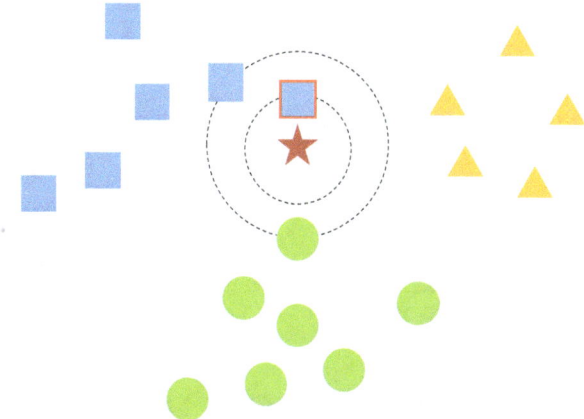

**Figure 2-4**   Classification process of the K-NN algorithm

## (2) SVM

In machine learning, SVM (Support-Vector Machines) are supervised learning models with associated learning algorithms that analyze data used for classification and regression analysis. Originally developed for binary classification, they are now widely applied to artificial intelligence. The reason for the popularity of SVM is that it provides the optimal hyperplane to solve problems in classification, so that the amount of data to be learned is easy to interpret, and also minimizes the empirical risk based on SRM (Structural Risk Minimization). The hyperplane is one dimension lower than the surrounding space. In three-dimensional space, hyperplanes are two-dimensional planes, and in two-dimensional planes, hyperplanes are one-dimensional lines. In the case of two-dimensional separation, the boundary is separated by a line, as shown in Figure 2-5 (a). In this way, the two-dimensional plane containing a mixture of black and white balls is divided into two parts, one containing the black balls, and one containing the white ones. Figure 2-5 (b) clarifies the border between the squares and the circles and gives the maximum margin to the border.

The hyperplane in Figure 2-5 is a line for classifying data. The support vector and the margin are used to find the optimal hyperplane to separate the two classes. There are two ways to obtain a hyperplane, the Hard Margin method and the Soft Margin method. The Hard Margin method is a very strict method of separating two groups, which is difficult to use when there is noise. The Soft Margin method uses a little slack on the boundary where the Support Vector is located.

Ultimately, in Figure 2-5, the final goal of SVM in Figure 2-5 is to find a split straight line that separates the two groups. The researcher should ensure that the hyperplane separates the two groups, but should position the hyperplane so that the two groups are as far apart as possible with the hyperplane in between. This is

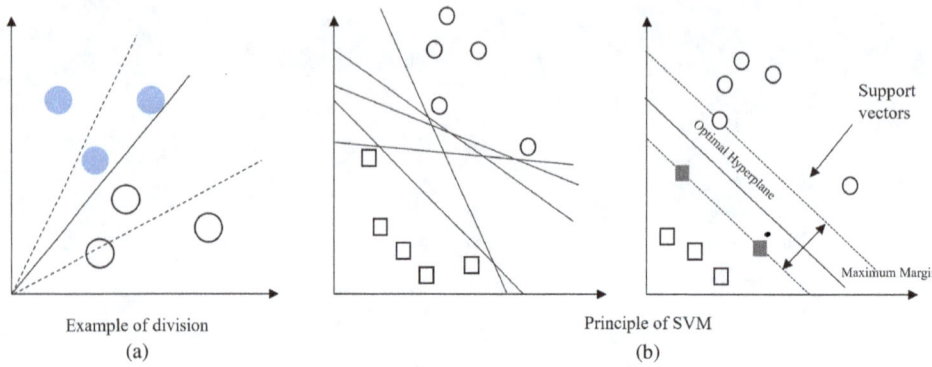

Example of division                    Principle of SVM
(a)                                          (b)

**Figure 2-5**   Division of SVM

because the farther apart the two groups are, the more reliably the group to which any new data belongs can be determined.

### (3) Decision Tree

Named as such because of its tree-like appearance, decision trees analyze the data points and present patterns that exist between them as a combination of predictable rules. It is a predictive modeling approach used in statistics, data mining and machine learning.

Decision trees can also be used for classification and regression. Decision trees are used when you want to find out which group the data belongs to. They work by dividing the data points into several groups with particular characteristics and then finding the characteristics of each group. It can be used to find rules which can then be used to predict future events.

## 2.3.2 *Regression*

Regression analysis is a set of statistical processes for estimating the relationships among variables. Regression infers the "function relationship", which determines which function generates the given data. Today, regression analysis is the most used statistical model as a result of its proven use in a variety of areas, including economics, medicine, and engineering.

Regression is used to express a hypothesis and is also used as a predictive model in supervised learning in machine learning. For example, as in the case of statements like "If you eat a lot, you gain weight fat," or "If the speed of the car is high, the probability of accidents is high", when there is a high degree of input, there will be a high degree of output.

## (1) Linear Regression

Figure 2-6 shows an example of linear regression. Linear regression is a regression technique that models the linear correlation of the dependent variable y with one or more independent variables (or explanatory variables) x.

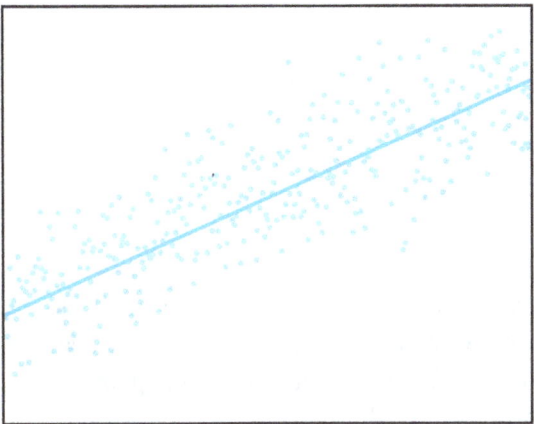

**Figure 2-6** Example of linear regression

For example, a company's sales can be expressed as through the independent variables of employee age, education and career. The dependent and independent variables must always be quantified. Linear regression analysis, which is used in statistics and data mining, is classified as a supervised learning model for learning with data sets in the field of machine learning. In statistics, data x is called an independent variable, and data y is called a dependent variable, while in machine learning, data x is called a characteristic, and data y is called a label. The process of obtaining arbitrary coefficients from these datasets is called regression estimation in statistics and learning in machine learning. After the training the predictive value for the new input data is predicted through the linear prediction model.

## (2) Logistic Regression

Figure 2-7 shows an example of logistic regression. Dependent variables of linear regression are generally used when they have a continuous normal distribution. If the dependent variable is expressed as categorical (yes/no, 1/0, pass/fail, buy/non-purchase), logistic regression is used. This is called the binary logistic regression model when the dependent variables fall into two categories.

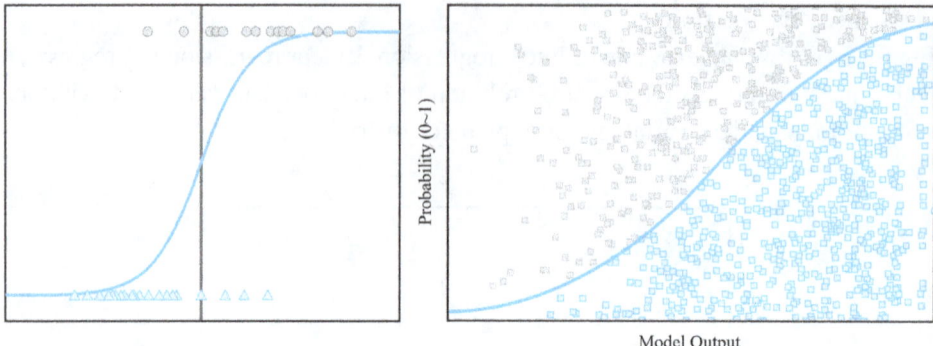

**Figure 2-7**   Example of logistic regression

Logistic regression, unlike linear regression analysis, can be viewed as a sorting technique because dependent variables are categorical data, and input data is divided into specific classifications. At this point, the agent proceeds with learning to maximize its reward.

### 2.3.3 *Reinforcement Learning*

Reinforcement learning is an algorithm that includes the process of collecting data in a dynamic environment. It can be considered as the algorithm that is most similar to the method by which a human learns because it learns about a certain situation through trial and error. The agent receives rewards from the environment by observing the conditions of the environment and taking appropriate action accordingly.

Figure 2-8 shows a series of processes that the agent performs that, together, results in the maximization of the reward received by the agent from the environment when the agent repeats the observation-action-reward interaction. The process of observation–action–reward undergone by the agent is also called experience. There are many fields of study that have advanced reinforcement learning, among them, behavioral psychology and control theory affected the greatest influence.

'Trial and error' in behavioral psychology is the principle of learning in humans and animals, which has been applied to machine learning. The agent learns to make the best decisions by remembering the rewards and penalties for all their actions. Another area of research that has had a significant impact on reinforcement learning is optimal control.

Optimal control, a theory that emerged in the late 1950s, involves designing a control device that optimizes the efficiency of a dynamic system. The dynamic system optimization problem is used to make optimal decisions for each process over time.

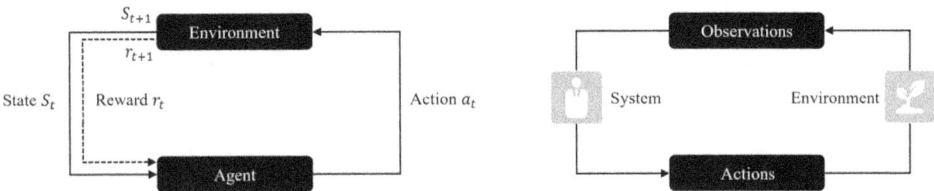

**Figure 2-8**　Types of reinforcement learning

American mathematician Richard Bellman solved discrete-time problems by introducing the Markov Decision Process (MDP)[2] model. Reinforcement learning is used to make decisions about sequential events because it takes place through feedback received as events unfold.

Some argue that the process of learning based on trial and error is very similar to the way people learn, and thus is the best model of artificial intelligence. Reinforcement learning models are most effectively used in games and robotics. In terms of machine learning, reinforcement learning can be considered as supervised learning using a control-based approach, as opposed to having a mathematical basis, but it is sometimes considered as unsupervised learning.

## 2.4 Unsupervised Learning

Unsupervised learning is a way of making a computer learn with unlabeled data in order to discover hidden features or structures in the data. The classification of supervised learning described previously is the creation of discriminants for classification based on a labeled training data set. These discriminants are later used to predict the label of new unknown data. On the other hand, unsupervised learning classifies unlabeled data, so its purpose is to group data sets into groups with the same characteristics. A representative model of unsupervised learning is clustering. A clustering model is that in which the training data does not have a label. In a clustering model, how the input data forms a group is important. In other words, the function of the clustering model is to analyze the characteristics of the data obtained without labels and to group data with similar characteristics. Except for the absence of labels, clustering models serve the same purpose as classification models of supervised learning. In other words, the clustering model classifies data without labels. A comparison of classification and clustering is shown in Figure 2-9.

---

[2]MDP is a stochastic process in which the future state of the system is determined irrespective of past history if only the present state is given. It is the probability of continuing to the current state (from the first state to the present) when certain S(state), A(Action).

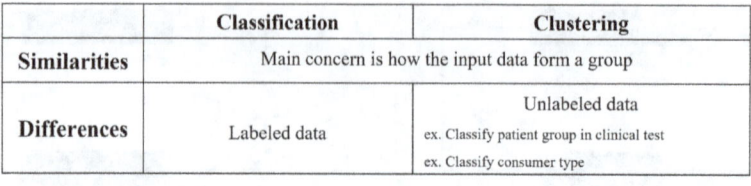

| | Classification | Clustering |
|---|---|---|
| **Similarities** | Main concern is how the input data form a group | |
| **Differences** | Labeled data | Unlabeled data<br>ex. Classify patient group in clinical test<br>ex. Classify consumer type |

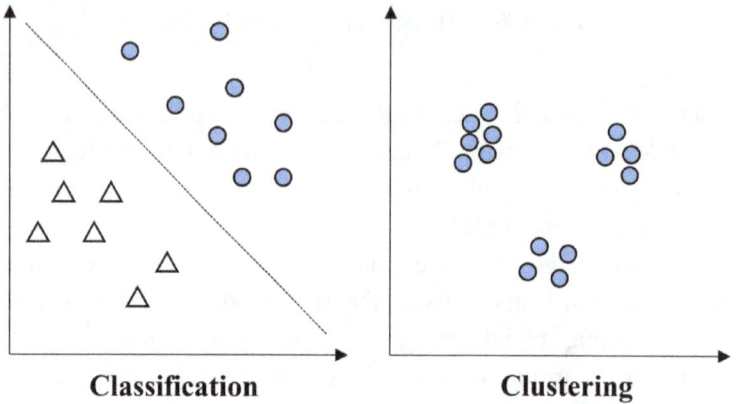

**Figure 2-9**    Classification model and clustering model

Clustering models can be largely divided into two models from the perspective of clustering algorithms. The first is the flat or partition-based clustering technique, and the second is the hierarchical clustering model. The k-means, k-medoids, and DBSCAN are representative of the partition-based cluster model.

Hierarchical clustering is distinguished by agglomerative clustering and divisive clustering. There are many different areas where the clustering model is applied. For example, differentiating human voices from noise to improve call quality on phones is a good example of a clustering model that distinguishes unlabeled data. In clinical tests in the medical field, the clustering model can be used to distinguish the DRG (diagnosis related group) or HG (hospitalists group) and can also be used to segment customers in marketing.

Recently, it has been used for market and customer segmentation in the field of marketing, while in social network services such as Facebook, it is also used to create communities for users with the same interests. Google uses the clustering model to group news of the same topic into the same category in its news service.

### (1) K-Means Clustering
The K-means algorithm is an algorithm that groups given data into k clusters. It is one of the simplest unsupervised learning algorithms. It groups unlabeled data in such a way as to minimize the variance of the distance difference between each

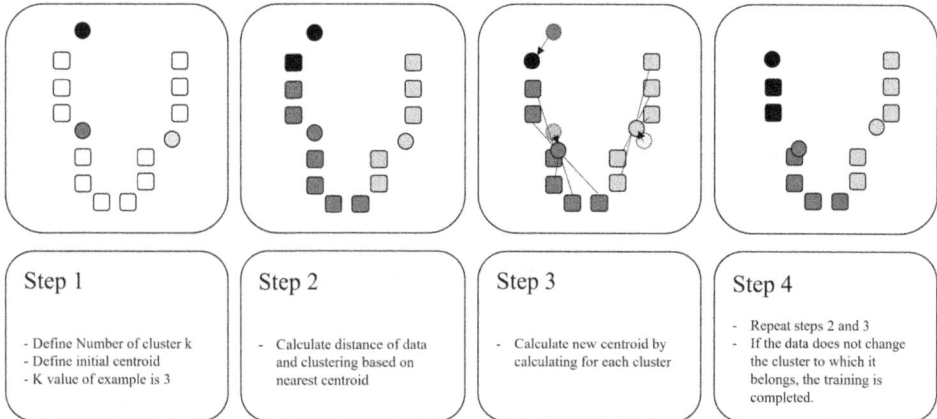

**Figure 2-10**  K-means clustering algorithm procedure

cluster. When using the K-means algorithm, we do not know how many clusters exist in the unlabeled data, so it is necessary to predetermine the number of clusters to classify. The k stands for the number of clusters.

After the number of clusters to classify is determined, the initial centroid of each cluster is initialized. The centroid used in the k-means algorithm is a hypothetical group representing a cluster, and each learning data point becomes the reference point for determining the cluster to which it belongs. The initial centroid is set arbitrarily, and as learning progresses, it moves to the optimal center for each cluster. Figure 2-10 shows the procedure for performing the K-means algorithm.

Finding the optimal centroid is the key to the k-means algorithm. The concept of finding the optimal centroid is very intuitive and simple. That is, after the distance of all data to k initial centroids are calculated, k being determined arbitrarily, the nearest centroid is considered as the centroid of the cluster to which they belong. Once k clusters are determined, the average of the coordinate values of the training data belonging to each cluster is obtained and set as the new centroid. The distance from all the training data to the k newly defined centroids is calculated, and the nearest centroid to each data point is defined as the cluster to which the data point belongs. These steps are repeated and terminate when the cluster members no longer change. The k-means algorithm is expressed as follows:

1. After determining the number of clusters (k = n), n centroids are set randomly.
2. The distance from all the data points respectively to n centroids is calculated and the nearest centroid to each data point is set as the cluster centroid to which that data point belongs.
3. The average of the coordinate values of the training data is calculated for each cluster and then set as the new centroid.

4. Steps 2 and 3 repeat until the cluster of each data no longer changes.
5. If there is no change in the cluster to which all the learning data belongs, learning is completed.

Like this, the k-means model is very fast at calculating data for learning. When the dimension of the training data is n, most other algorithms have $O(n^2)$ and $O(n\log n)$ computations, where k-means is $O(n)$. Consequently, the complexity of a program in software engineering is calculated by calculating the complexity of the algorithm with the least complexity.

### (2) DBSCAN

Like the k-means algorithm, DBSCAN <sup>Density based spatial clustering of application with noise</sup> is an unsupervised learning model and, as the name suggests, is a cluster model that is strong in identifying noise and outlier data.

DBSCAN is the density-based clustering algorithm and is the most used algorithm.

When clusters are produced, data outside of the dense region is considered as noise or the boundary point.

In the K-MEANS model, the distance from all the data points to the K centroids is calculated to determine which cluster each data point belongs to and which cluster center is the shortest distance from each data point, but as shown in Figure 2-11, the DBSCAN model determines that data groups with a

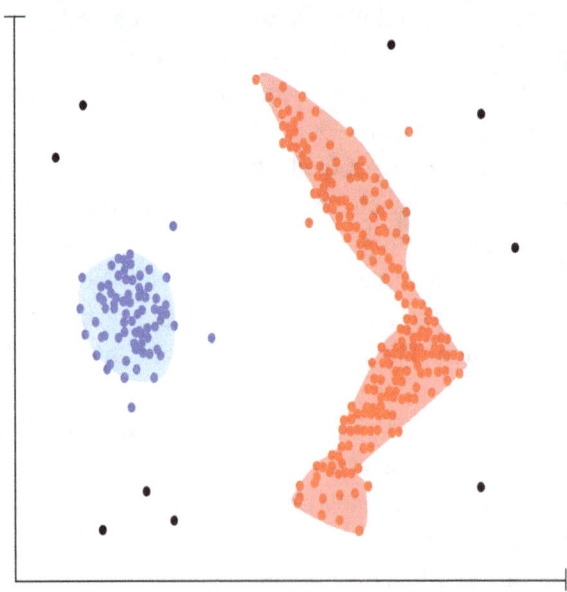

**Figure 2-11**   Example of density-based clustering algorithm

certain density are considered as the same cluster regardless of the concept of distance.

## 2.5 The Difference Between How Machine Learning and Statistics Work

Many researchers may be familiar with regression analysis of statistics.

Ordinary least squares estimation, a representative method of multiple linear regression analysis in statistics, finds the regression coefficients by minimizing the distance between each observation and hyperplane and verifies their statistical significance.

Machine learning also uses the least squares estimation for regression analysis but advances in computing have led to a preference for gradient descent that can do rapid analysis.

The gradient descent method selects an arbitrary weight on the error surface drawn by a combination of weights[3]. It then proceeds by input-learning the data and updating the weights until the minimum error on the error surface is reached.

Figure 2-12 is a graph showing the path from the error surface to the minimum error value according to learning.

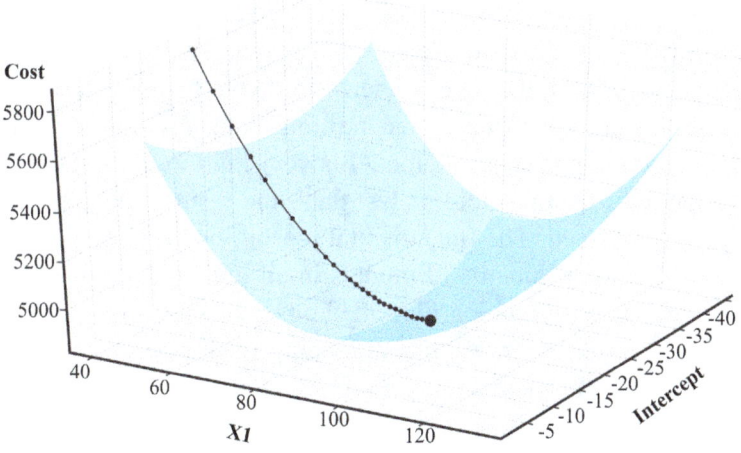

**Figure 2-12**   A graph showing the path from the error surface to the minimum error value according to learning

---

[3] In the linear regression analysis equation of machine learning, $H(x) = Wx + b$, for the W corresponding to the slope of the independent variable x. It is expressed as a regression coefficient in Statistics, but as a weight in machine learning.

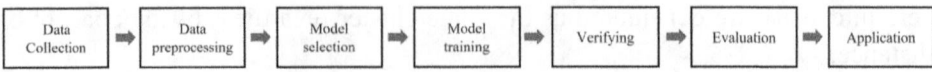

**Figure 2-13**   Machine learning procedure

The discipline of statistics verifies statistical significance and examines the relationship between variables in order to estimate the population using given data.

Machine learning, on the other hand, learns inductively based on data and modifies weights. Since there is no verification process for determining statistical significance, in order to establish whether the result of machine learning is appropriate, performance evaluation must be carried out through the construction of a set composed of training data, verification data and test data. Figure 2-13 shows the machine learning procedure.

## 2.6  Considerations for Performing Machine Learning

It is important to consider the problems of underfitting and overfitting in analysis using machine learning. Underfitting means that the model is simple, so it cannot adequately explain the input data and is thus less predictive. Whether the model is underfitted or not can be judged according to the cost, which is the error between the model's predicted value and its actual value. Model selection and the obtainment of additional training data may compensate, to some extent, in the case of underfitting. Overfitting is a case in which the model is too closely fitted to the training data. As shown in the figure below, the predictive power of the model is very high in the training data, but when it is applied to actual field or test data, the result is that the predictive power is greatly reduced. The fundamental reason for overfitting is that major machine learning algorithms build models in an inductive manner. If you are analyzing using the machine learning method, you will have to go through a series of model validation procedures, keeping in mind that such overfitting problems can occur at any time.

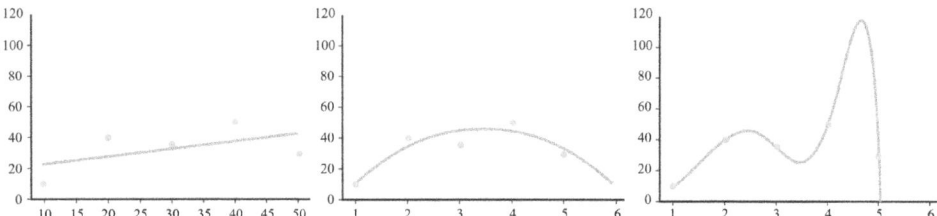

**Figure 2-14**  Examples of underfitting, appropriate fitting, and overfitting

Figure 2-14 shows various types of learning statuses.

## 2.7  Resources for Machine Learning

This chapter introduces the data sources available for machine learning analysis.

### 2.7.1 *Kaggle*

The address https://www.kaggle.com is the Kaggle site address. Kaggle is a site for data science and machine learning. Kaggle hosts competitions in which data scientists and global companies compete to find out who can classify or predict most efficiently. Competition topics vary and change all the time. Because winners' algorithms are featured on the site, it is a good resource for researchers interested in machine learning.

　　If the topic is of interest to the researcher and the data is difficult to obtain, the data uploaded to Kaggle can be pre-tested to guide the overall direction and determine the progress of the research. For each competition it hosts, Kaggle uploads a competition data set and a variety of data including text, images, and sound for researchers to test and analyze. Figure 2-15 shows part of Kaggle data set list.

**Figure 2-15**    Part of the Kaggle data set list

(https://www.kaggle.com/datasets)

## 2.7.2 *Public Data Portal*

The public data portal is a site that integrates and manages public data created and managed by public institutions. In addition, public data can be easily accessed through the OPEN API. (https://www.data.go.kr)

If data is held by public institutions, but cannot be found in the public data portal, you can contact the public institutions in question or the relevant ministries to make such data available for download.

Even if you do not use a public data portal, secondary data are provided directly by public institution sites such as Statistics Korea and the Korean Center for Disease Control and Prevention. Public data can be downloaded and used once application documents and a fee have been submitted. Representative secondary data that can be downloaded from each institution are as follows. In addition to the sites listed in the table below, various organizations in each field also release secondary data. Table 2-1 shows representative secondary data.

**Table 2-1**   Representative secondary data

| Data | Theme | Organization | Website address |
|---|---|---|---|
| National Health Insurance Service Big Data | Health and medical treatment | National Health Insurance Service (NHIS) | https://nhiss.nhis.or.kr/bd/ay/bdaya001iv.do |
| National Health and Nutrition Examination Survey | Health and medical treatment | Korea Centers for Disease Control and prevention (KCDC) | https://knhanes.cdc.go.kr/ |
| Youth Health Behavior Online Survey | Health and medical treatment, Youth | Korea Centers for Disease Control and prevention (KCDC) | https://yhs.cdc.go.kr/ |
| Korea Welfare Panel | Welfare, Society | Korea Institute for Health and Social Affairs (KIHASA) | https://www.koweps.re.kr:442/ |
| Korea Labor and Income Panel | Labor | Korea Labor Institute (KLI) | https://www.kli.re.kr/klips/index.do |
| Workplace Panel Survey | Management | Korea Labor Institute (KLI) | https://www.kli.re.kr/wps/index.do |
| Korean Longitudinal Survey of Women and Families | Women, Society | Korea Women's Development Institute (KWDI) | http://klowf.kwdi.re.kr/ |
| Statistical Geographic Information Service | Geographic information | Statistics Korea (KOSTAT) | https://sgis.kostat.go.kr/view/index |
| Seoul Survey | Society | Seoul Metropolitan Government | http://data.si.re.kr/sisurvey |

## 2.8  Practice Questions

Q1. What is machine learning?

Q2. Explain the difference between machine learning and data mining.

Q3. Describe supervised learning and unsupervised learning.

Q4. Describe the methods used in supervised learning and unsupervised learning.

# Chapter 3

# Deep Learning

## 3.1 Definition and Concepts of Deep Learning

Artificial intelligence is defined as the understanding of human judgment, action, and cognition which we draw upon to equip machines[1] with human intelligence. Recently, there seems to be a lot of confusion about machine learning and deep learning. Therefore, simply put, artificial intelligence is a prerequisite for a machine to possess human intelligence, and it must "learn" in order to produce this artificial intelligence. One class of this kind of learning is machine learning, while deep learning is the class of learning similar to neural networks for machine learning. Figure 3-1 shows these relationships.

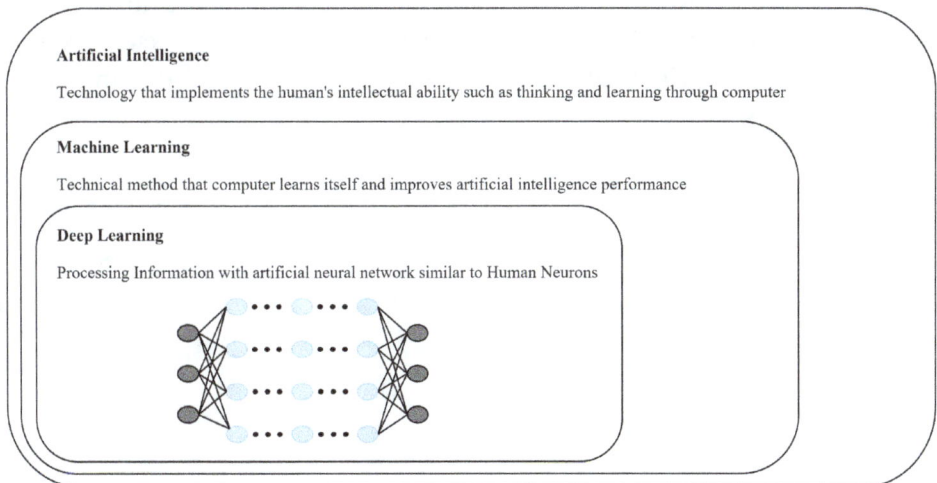

**Figure 3-1**　Diagram demonstrating the relationship between artificial intelligence, machine learning and deep learning

---

[1] In artificial intelligence, machines refer to microcomputers, PCs, servers, etc.

In short, deep learning is a subset of machine learning that the way humans think to a computer. It allows a computer to learn on its own, and is based on a high-level abstraction[2] that represents the human brain with artificial neural networks. Neurons, nerve cells in the brain, are unique cells that carry electrical signals. The brain is a collection of neurons, with about 100 billion cells and 100 trillion connections. These neurons are connected to each other, and the resulting networks are called neural networks, while artificially created ones are called artificial neural networks. Figure 3-2 shows a neural network with neurons.

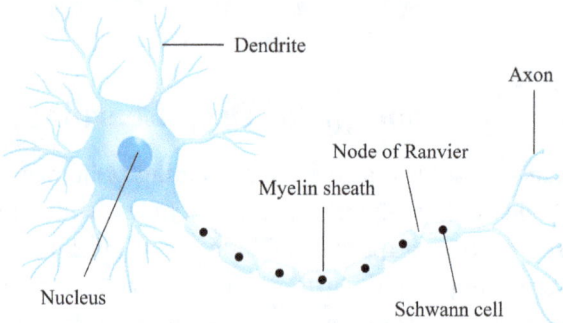

**Figure 3-2**   Neuron

Neurons receive signals from dendrites and transmit signals from axons, and there are synapses between neurons. Only signals higher than or equal to the threshold are transmitted, and information is transmitted or stored in the electrical signal.

The flow of artificial intelligence is expressed in mathematical terms and in terms of neurons, as shown in Figure 3-3.

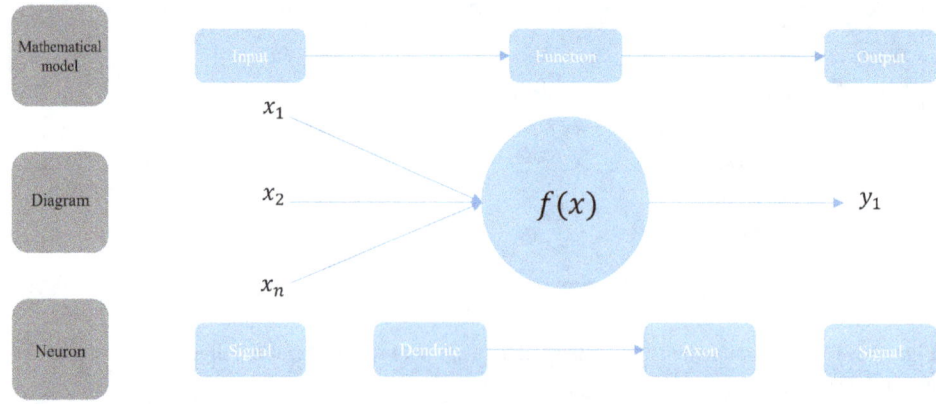

**Figure 3-3**   Example of neural network modeling

---

[2] An abstraction summarizes key content or functions of large amounts of data or complex data

Deep learning technology is already widely used by global IT companies such as Google, Facebook, Amazon, and is known to be particularly effective in natural language processing, such as pattern recognition, photo recognition, voice recognition, and machine translation.

Deep learning is divided into perceptron, multilayer perceptron, and current deep learning models.

### 3.1.1 *Perceptron*

The origin of the neural network can be traced to the perceptron proposed by Frank Rosenblatt in 1958. The concept behind the perceptron is that each input is weighted for n inputs, and one output is produced. This is how neurons transmit information as electrical signals. That is, for a model of supervised learning using labeled training data, the training data adjusts the weight of each input until the predicted value is equal to the actual result value corresponding to the characteristic value. When a neuron receives an input signal, it transmits the signal using the dendrites and axons, and the perceptron weights the input. Figure 3-4 shows the concept of the perceptron.

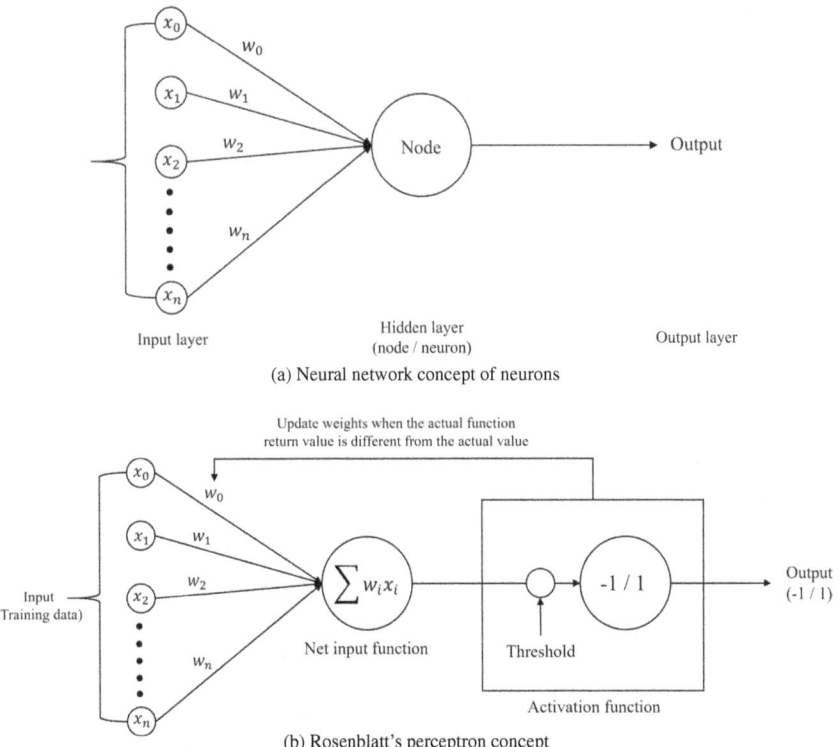

(a) Neural network concept of neurons

(b) Rosenblatt's perceptron concept

**Figure 3-4**   The concept of the perceptron

The difference between (a) and (b) of Figure 3-4 is that in (b), a weight is assigned to each of the n input signals. The perceptron outputs 1 when the sum of the signals after the calculation exceeds the threshold, or 0 or -1 otherwise.

Since the weights divide the boundaries of the thresholds, the weights serve as parameters for how much the inputs will affect the outputs. As such, this can be classified as a linear classification model because it distinguishes cases in which a value is within the threshold and cases in which it is not, and therefore it can also be regarded as a form of supervised learning.

In other words, the input values or neurons are entered into a net input function, weighted, and made into a single value. A net input function is defined as a function that combines the weighted inputs and outputs '1' if the result is greater than the threshold, or otherwise outputs '-1'(Figure 3-4). The perceptron marks the beginning of artificial neural networks, but as with all beginnings, the perceptron began to show limitations in solving very simple XOR models. Therefore, the multilayer perceptron would emerge.

### 3.1.2 *Multilayer Perceptron*

The multilayer perceptron was introduced so that the disadvantages of the perceptron could be overcome. The perceptron was unable to learn something as simple as XOR, but with the multilayer perceptron, the solution would prove to be surprisingly simple. A multilayer perceptron has several layers, not just one. Therefore, it has many intermediate layers called hidden layers, and the more hidden layers there are, the greater the improvement in classification power. Figure 3-5 shows the concept of multilayer perceptrons.

The introduction of multiple perceptrons solved the problem, but as the number of hidden layers increased, the weight also increased, and thus learning became difficult. To solve this problem, the backpropagation method, similar to the feedback control theory of control theory, was applied to enable multilayer perceptron learning.

In the field of deep learning, continuous research was conducted with these various methodologies, and new artificial neural networks appeared as the old control theory-based learning methods were re-examined. The Boltzmann machine[3], an unsupervised learning method, was proposed and attracted the attention of the academic world.

---

[3] A stochastic recurrent neural network

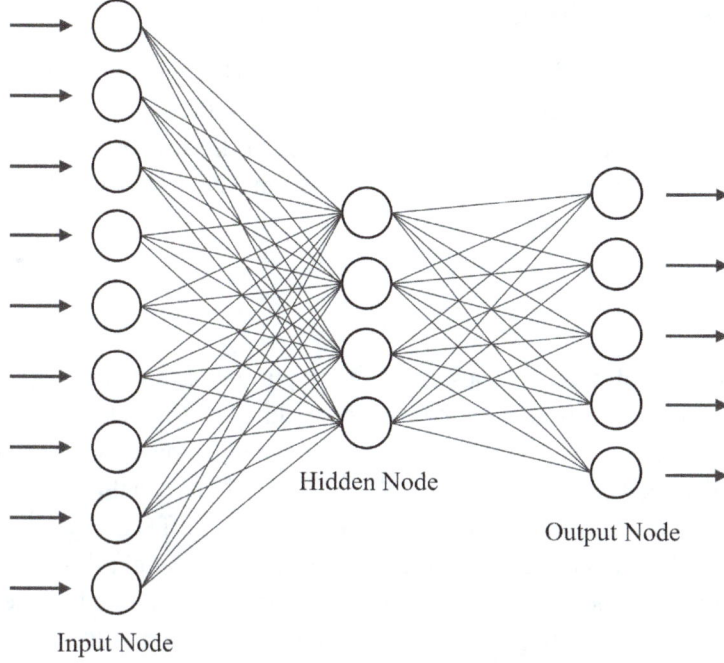

**Figure 3-5**   Concept diagram of a multilayer perceptron

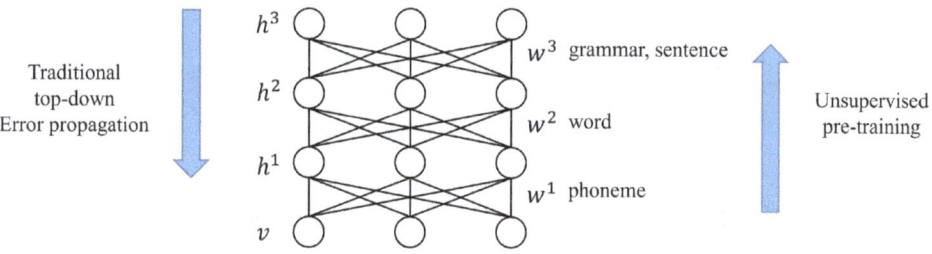

**Figure 3-6**   Expression of unsupervised learning

Figure 3-6 shows, Hinton and Terry conducted a sufficient degree of learning with unlabeled data beforehand, and then conducted supervised learning.

This approach solves many of the drawbacks of multilayer perceptrons, and many problems can be solved by conducting unsupervised pre-training using unlabeled data.

## 3.2  Types of Artificial Neural Network

There are many types of artificial neural networks. These are made up of multiple complex inputs and directional feedback loops (unidirectional or bidirectional) and various layers. However, artificial neural networks are generally divided into DNN (Deep Neural Networks), CNN (Convolutional Neural Networks), and RNN (Recurrent Neural Networks).

### 3.2.1  *DNN*

An artificial neural network consists of one input and one output layer from the perspective of a perceptron, with at least one middle layer in between. However, in DNNs, there are multiple hidden layers between the input layer and the output layer.

Basically, the DNN is an RBM[4 Restricted Boltzmann Machine], a model in which the interlayer connection is removed from Boltzmann machines. If intralayer connections are removed, RBM is a method that calculates the final weight through a process of fine-tuning, after some correction through prior learning, based on the shape of an undirected bipartite graph form consisting of visible and hidden layers. Thus, DNN is an applicable method even if there are not enough labeled data sets.

Figure 3-7 shows the DNN model.

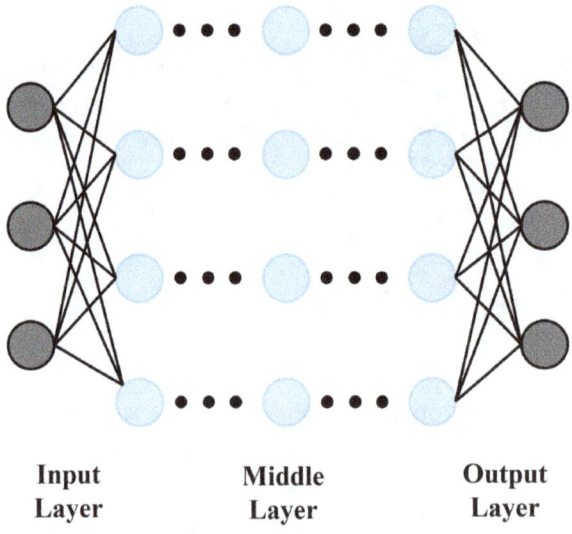

**Input Layer**    **Middle Layer**    **Output Layer**

**Figure 3-7**   Common DNN Model

---

[4] A Restricted Boltzmann Machine (RBM) is a model proposed by Geoff Hinton as an algorithm that can be used for dimension reduction, classification, linear regression, collaborative filtering, feature learning, and topic modeling.

Deep neural networks are used to adjust individual weights to output the desired values for an input layer. This can be learned with backpropagation algorithms widely used in machine learning, with the advantage of being able to obtain slow but stable results.

### 3.2.2 *CNN*

A CNN <sup>Convolutional Neural Network</sup> is a type of multilayer perceptron designed to use minimal preprocessing, consisting of one or several convolutional layers and a common artificial neural network layer on top of it. Thanks to this preprocessing structure, good performance is achieved in both video and audio fields compared to other deep learning structures.

Figure 3-8 shows the learning process of CNN.

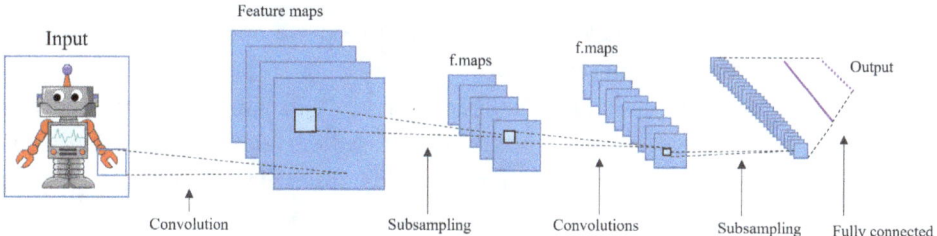

**Figure 3-8**   Learning process of CNN

In recent years, CDBNs (Convolutional Deep Belief Networks) have been developed in deep learning. They are very similar in structure to the existing CNNs, and are frequently used for image processing and object recognition because the two-dimensional structure of pictures can be analysed effectively by CDBNs. Some CDBNs are pre-trained using DBNs (Deep Belief Networks). A CDBN provides a general structure that can be used for various image and signal processing techniques.

### 3.2.3 *RNN*

One feature of an RNN <sup>Recurrent Neural Network</sup> is that the unit connection has a recursive structure so that it can store incoming input data and the current internal state of the neural network, while at the same time considering past input data. Therefore, by storing the state inside the neural network to model time-varying dynamic features so that it can process the input of the sequence type using internal memory, processing data with time-varying features, such as handwriting recognition

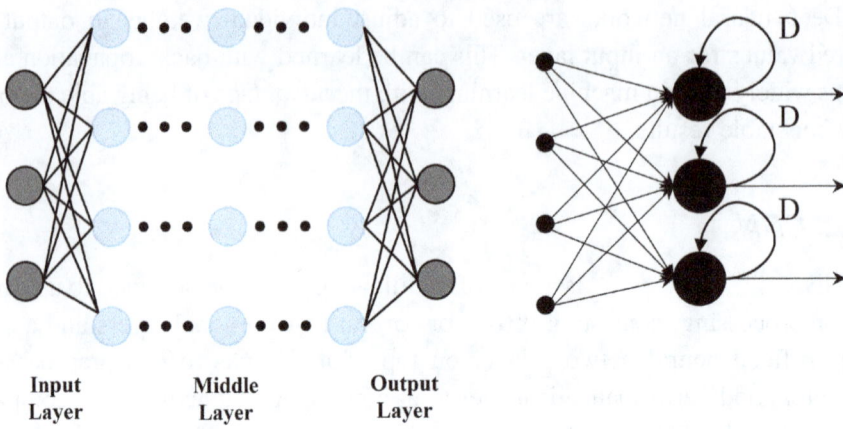

Input          Middle          Output
Layer          Layer           Layer

**Figure 3-9**   Comparison of neural networks and RNN networks

or speech recognition, is made easy. Figure 3-9 shows a comparison of existing neural networks and RNN networks.

As shown in Figure 3-9, the RNN has a structure in which a specific part of the neural network is repeated, recurrent. Gradient descent, Hessian Free Optimization, and Global Optimization Methods are typically used to train the RNN.

# 3.3 Practice Questions

Q1. What is deep learning?

Q2. What is a perceptron?

Q3. Describe the types of artificial neural networks.

# Chapter 4

# Case Study

## 4.1 AlphaGo

AlphaGo is an artificial intelligence Go program developed by Google's DeepMind Technologies Limited.

In October 2015, AlphaGo won all five times against Fan Hui of France, the three-time winner of the European Go Championships, becoming the first computer Go program to beat a professional Go player without a handicap. In March 2016, it beat Lee Sedol in a five-game match, the first time a computer Go program has beaten a 9-dan professional without a handicap. Although it lost to Lee Sedol in the fourth game, Lee resigned in the final game, giving a final score of 4 games to 1 in favor of AlphaGo. AlphaGo was awarded an honorary 9-dan by the Korea Baduk Association. In addition, since it is registered as a guest competitor in Korea, it may participate in competitions held in Korea at some point, although this is unlikely. Figure 4-1 shows the AlphaGo system that was used.

Go is considered much more difficult for computers to win than other games such as chess, because its much larger branching factor makes it prohibitively difficult to use traditional AI methods such as alpha–beta pruning, tree traversal and heuristic search.

However, while data and computational power were essential to AlphaGo's victory, algorithms for AI learning were the more decisive factor. The key was to reduce the number of cases, of which there originally are close to infinity. As such, AlphaGo was designed to allow trained DNN (Deep Neural Networks) to make the most advantageous choices through Monte Carlo Tree Search (MCTS). The system's neural networks were initially bootstrapped from human gameplay expertise. AlphaGo was initially trained to mimic human play by attempting to match the moves of expert players from recorded historical games, using a database of around 30 million movements. By applying this technology, AlphaGo could beat humans at Go.

**Figure 4-1**    AlphaGo system

(Source: https://3.bp.blogspot.com/-_-DoMQ0CuAQ/VzzPgbjhjOI/AAAAAAAACp4/5v6prECMLigEfJ23Vigp
jPWXoH-IT05UgCLcB/s640/tpu-1.png)

### 4.1.1 *System Configuration*

Initially, at the time of development, it was announced that the hardware would use parallel computations using CPUs and NVIDIA GPUs. AlphaGo has two versions: a single version, which runs as a single computer, and a distributed version, which uses multiple computers connected to a network. The single

**Table 4-1**  Configuration and performance

| Configuration | Search threads | No. of CPU | No. of GPU | Elo rating |
|---|---|---|---|---|
| Single | 40 | 48 | 1 | 2,181 |
| Single | 40 | 48 | 2 | 2,738 |
| Single | 40 | 48 | 4 | 2,850 |
| Single | 40 | 48 | 8 | 2,890 |
| Distributed | 12 | 428 | 64 | 2,937 |
| Distributed | 24 | 764 | 112 | 3,079 |
| Distributed | 40 | 1,202 | 176 | 3,140 |
| Distributed | 64 | 1,920 | 280 | 3,168 |

version of AlphaGo operates with 48 CPUs and 4-8 GPUs and has recorded only one loss in 500 games (including Crazy Stone and Zen). The distributed version consists of 1,202-1,920 CPUs and 176-280 GPUs. An early version of AlphaGo was tested on hardware with various numbers of CPUs and GPUs, running in asynchronous or distributed mode. Two seconds of thinking time was given to each move. The resulting Elo ratings[1] are listed below. Table 4-1 shows configuration and performance of early version of AlphaGo.

### –  AlphaGo Fan

In October 2015, the distributed version of AlphaGo defeated the European Go champion Fan Hui, a 2-dan (9-dan being the highest) professional, 5-0. In the battle with Fan Hui, AlphaGo used 1202 CPUs and 176 GPUs. This is known as AlphaGo Fan, Version 12.

### –  AlphaGo Lee

In March 2016, Lee Sedol's matchup was a distributed version that used 48 TPUs instead of GPUs. At that time, only the improved Version 18 of machine learning was known. However, in May 2016, Google unveiled its proprietary hardware "Tensor Processing Units," and stated that it had already been deployed in multiple internal projects at Google, including the AlphaGo match against Lee Sedol. Figure 4-2 shows a picture of the TPU.

---

[1] The Elo rating system is a method for calculating the relative skill levels of players in zero-sum games such as chess.

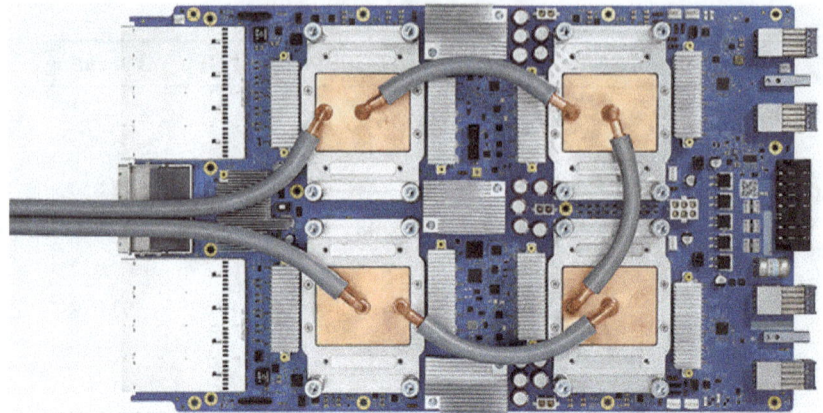

**Figure 4-2**   Tensor processing unit

(Source: https://cloud.google.com/images/products/tpu/cloud-tpu-v3-alpha-2x.png)

– **AlphaGo Master**

AlphaGo Master is a single version with four TPUs. It won 60 consecutive games against a professional Go player in early 2017 and won against Ke Jie in May of the same year.

AlphaGo Master, powered by one second-generation TPU module, was unveiled at the Google I/O 2017 conference on May 17, 2017.

The TPU module used here is composed of four TPUs with 45TFLOPS (teraFLOPS, 45 trillion operations per second), and 180TFLOPS performance, with one module supporting 64GB memory bandwidth. Google explained that the computational performance of the TPU is 30 to 80 times higher than the latest CPU at the time.

Based on what the previous AlphaGo learned, AlphaGo Master can learn at the same time as it engages in reasoning, and the time required for learning has been reduced to one third. The physical volume of the machine has also been reduced, resulting in a 10-fold increase in energy efficiency.

– **AlphaGo Zero**

This single version with four TPUs is the final version of AlphaGo. It was introduced on October 19, 2017, in "Nature", through the presentation of a thesis titled "Mastering the Game of Go without Human Knowledge."

AlphaGo Zero learns and improves itself utilizing Go rules without supervised learning that depends on human methods. It surpassed AlphaGo Lee in 36 hours of learning, lost one in 100 matches against AlphaGo Lee after 72 hours of learning, and recorded 89 wins and 11 losses in a match against AlphaGo Master 40 days later. During this period, AlphaGo Zero learned from 29 million different matches. The emergence of artificial intelligence, which does not require the learning of big data, is meaningful in that it offers solutions to areas in which it is difficult to utilize artificial intelligence because of the difficulty of obtaining big data (unlike the case of Go).

**– Alpha Zero**

AlphaZero is a generalized artificial intelligence program that can be applied to board games such as Go, chess, and others as an algorithm. It took two hours for Alpha Zero to beat Elmo, the 2017 CSA world champion, at Shogi; four hours to beat Stockfish, the 2016 champion, at Chess; and 30 hours to beat AlphaGo Zero at Go.

On December 14, 2017 (U.S. time), Dr. Aja Huang (Staff Research Scientist at DeepMind, DeepMind Technologies Limited) announced that she would end her journey with AlphaGo and shift over all of AlphaGo's resources to other artificial intelligence development projects.

## 4.1.2 *Algorithm Implementation*

Until AlphaGo appeared, Go was believed to be the only analog game that a computer could not approach from the perspective of a human being. Artificial intelligence-based Go, which had been developed up until that point, was far too different from that played by professional Go players. Therefore, what kind of method did AlphaGo use to distance itself from previously existing algorithms and to beat the world's top professional Go players?

Let's take a closer look at the structure of the game of Go.

**– AI Approach Strategy**

Let's look at how AI approaches problems in a 1-on-1 board game. Suppose we have been asked to create artificial intelligence that wins board games without considering a game's time limit, when a game could potentially take an infinite amount of time.

As there would be no time limit, an ideal method would be to look as far ahead as possible and take the ideal course of action in the present state.

In the end, one would have to predict what the other person will do in relation to one's own move. Among the algorithms that could be used to solve this problem, the Minimax Algorithm is one of the most representative. Figure 4-3 shows a tree diagram to explain the Minimax Algorithm.

The circles represent the moves of the player running the algorithm, while the squares represent the moves of the opponent. When it considers all the possible moves of the opponent and the player running the algorithm, the scores that the latter could obtain after four moves are marked on the 4 layers. The highest

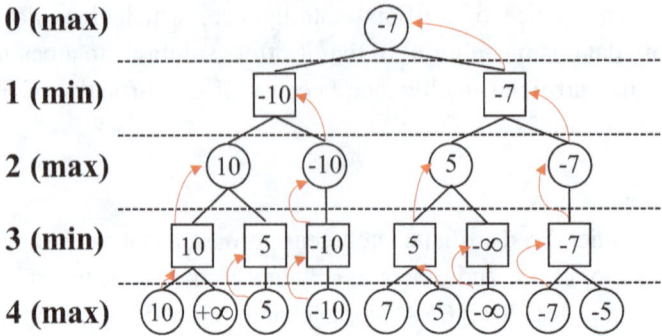

**Figure 4-3**   Minimax algorithm

score attainable by the player running the algorithm is + infinity, which would clearly be the most desirable outcome for that player.

However, of course, if the opponent is a reasonable player, he will act in a way that minimizes the points the player running the algorithm can earn. Therefore, the player running the algorithm will be acting to get a smaller +10 score of +10 or + infinity. Also, in their turn, both the player running the algorithm and the opponent will choose to maximize their score.

With this in mind, when one moves from level 1 to level 0, one should neither choose to get the highest score of +10 or + infinity points as shown on the left-hand side, but rather, when considering the moves of the opponent, the best course of action would be to go for the score of '-7' on the far right-hand side.

In a one-on-one game like Go, this behavior is known as the ideal choice, and the minimax algorithm is applied in the sense that you must choose the best of the worst numbers your opponent provides, thus minimizing possible loss. In the end, creating a powerful AI requires a new algorithm that efficiently navigates these search spaces, looking for behaviors that use less time and perform better than the minimax algorithm.

Various algorithms such as alpha-beta pruning and principal variation search have been suggested to find this method. The following method is the Monte Carlo Tree Search (MCTS), which is the most widely used AI for board games and is based on AlphaGo.

**–   Monte Carlo Tree Search (MCTS)**

In computer science, the Monte Carlo tree search (MCTS) is a heuristic search algorithm for some kinds of decision processes, most notably those employed in game play. It is a generic term for "probable ways of solving problems." This approach uses randomness to approximate the elaborate formulas that are hard to find, rather than huge search spaces that are hard to see. For example, suppose you

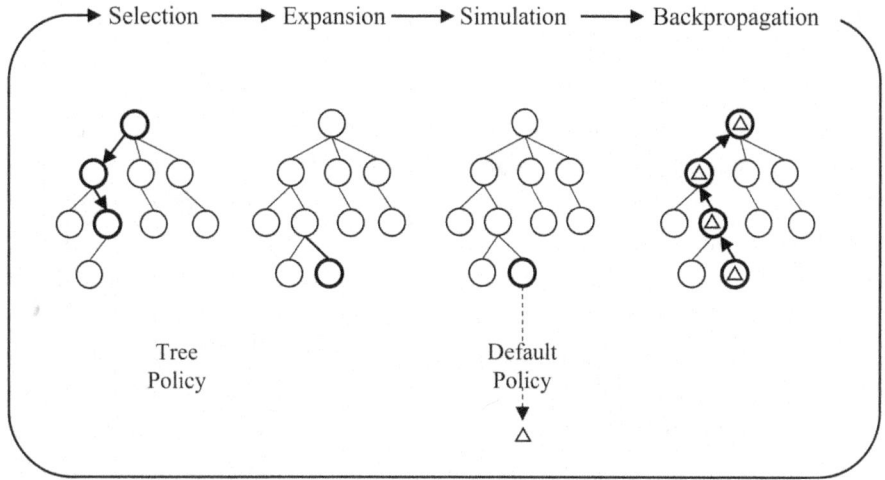

**Figure 4-4**   Monte Carlo Tree search diagram

draw a circle with a radius of 1 inscribed on a 2 x 2 square, and then randomly draw a point in this square to calculate the number of points in the circle / the total number of points taken. After that, if you draw many points, the value will eventually converge on the value of $\pi/40$.

This is an example of a typical Monte Carlo method, using Monte Carlo to calculate the value of the circumference. Monte Carlo Tree Search is an algorithm designed to select a good course of action without needlessly searching for all possible states by applying the Monte Carlo technique to the behavior tree search using the aforementioned minimax algorithm.

Let's look at the most basic form of MCTS, and how MCTS uses randomness to construct a tree.

The Monte Carlo Tree Search Diagram is shown in Figure 4-4.

There are four basic steps in MCTS.

- Selection: Start from the root and move to the Leaf Node by picking the most successful child node among the other child nodes. In general, to avoid searching on only one side, MCTS is calibrated in such a way as to give options randomly.
- Expansion: If a leaf node is reached, and it is not the final node at the end of the game, then create a child node for the next step, choose one of the child nodes among them, and move.
- Simulation: Proceed with playing the game by performing random simulations in the current state. This is called playout, and one progresses through the game until the outcome of the game is decided (for example, whether you win or lose).

- Backpropagation: It updates the weights on the nodes of the paths based on the results of the current simulation.

MCTS repeats this step as many times as possible to simulate information regarding the winning rate of each action that I might take

When doing a degree of searching and then finally selecting an action, one should choose the action that has been simulated with the largest number of possible actions on the route. Surprisingly, it has been proved that if the MCTS is repeated indefinitely, the result converges with the result of minimax.

However, in general, the convergence speed of basic MCTS is very slow, so there is a need for a method that reduces search space and finds results more efficiently.

The final result that MCTS wishes to obtain from converging is the probability of winning on each node (game status) in the tree. In this regard, suppose that there is a theoretical function v * (s) that calculates the degree of probability that I can win the game in this state when the current state "s" of the game is given. Of course, we do not know what these functions look like and how they work, but we can assume the existence of these functions. When the MCTS proceeds indefinitely, the values of each node of the MCTS converge to value function v *(s) — in each state.

Conversely, if we have a function v(s) that can approximate v*(s), because we can first get the approximate results we want from MCTS, we may be able to calculate our ideal move, even without proceeding with MCTS too deeply. In other words, creating function v(s) that calculates the win rate from the current state "s" of the game is important.

Previous researchers utilized MCTs and v(s) as methodological approaches, and used them to create Go AI, such as Pachi and Fuego.

However, because function v(s) has proven too difficult to make, previous studies have failed to create powerful v(s) functions, which in turn has led to a weakening of AI performance. At this point in time, however, Google's DeepMind team has greatly improved performance by transplanting deep learning into the process of creating v(s).

The focus of MCTS is on the analysis of the most promising moves, expanding the search tree based on random sampling of the search space. The application of Monte Carlo tree search in games is based on many playouts, also called rollouts. In each playout, the game is played out to the very end by selecting moves at random. The final game result of each playout is then used to weight the nodes in the game tree so that better nodes are more likely to be chosen in future playouts.

– **AlphaGo Pipeline Structure**

So far, we have learned that AlphaGo expertly navigates the state-action tree using the MCTS technique and the state evaluation function v (s) learned using deep learning. In reality however, the algorithm is computed in a way that is much more complicated than the explanations previously looked at up until this point, and, rather than one neural network there are several consisting of complementary structures.

Let's first look at how the neural networks inside AlphaGo are organized.

First, AlphaGo has a network that "predicts the location of the next stone in the current state" (that is, it not only predicts where the opponent will put the stone in the current state, but also predicts where I should put it). In certain papers, this is called a 'policy network'. Naturally, this policy network takes the form of a CNN, so AlphaGo first needed a large amount of data. As stated previously, a CNN requires a great deal of input data and the corresponding output data. To get the output data, we used Go game records from the website "KGS".

Using these game records, a total of 30 million pieces of data were given about where to place the next playing stone, and these were used to learn the policy network. In addition, we prepared and made AlphaGo learn a network called 'Rollout policy,' which can calculate the same problem much faster than policy networks.

Policy networks are also highly effective networks. The policy network was able to win Go games against high odds without using MCTS. However, networks alone cannot power AlphaGo. As we saw earlier, the search for MCTS required a "value network," a v(s) that calculates whether the player can win in the current state. Value networks have CNN structure similar to policy networks.

However, unlike policy networks that compute probability distributions from the last position of a stone, value networks yield only one simple value for state S, given as input at the end. To learn this network, AlphaGo saved the Go data records from self-play games, collected the winning and losing states, put them into the value network's learning data, and successfully completed the learning. As such, it is believed that AlphaGo conducted MCTS using Policy Networks, Rollout Policy, and Value Networks.

Each network is used in the following manner.

- Policy Network: In the original selection phase, MCTS uses the method of selecting the most successful child node.

However, AlphaGo's MCTS uses a method that adds a term for "the probability of releasing this stone in its current state", calculated by the policy network. This causes it to move more towards a higher probability action. This can be interpreted as AlphaGo implementing a form of human intuition that could not be implemented by a conventional computer algorithm.

MCTS also encourages players to search for actions in the other direction by giving them a certain amount of penalties if a player visits only one side. This is to prevent a search from becoming too one-sided.

- Rollout Policy: Rollout policy has much lower performance than a policy network but it does have the advantage of being able to calculate much faster. Therefore, when running simulations in the Evaluation stage of MCTS, a rollout policy network is used.
- Value Network: Value networks are also used in the evaluation phase. In the evaluation stage, both rollout policy and a value network calculate the scores independently, and the final scores of this state are calculated by adding the two at a specific ratio. According a number of papers, it is much better to add the scores by using a combination of the two (ELO Rating 2900) rather than using only either (ELO Rating 2450 and 2200 in a single machine) rollout policy or a value network.

After completing the selection, expansion, and evaluation process for each step, the score $v(s)$ evaluated for the selected state is displayed. This score is used to update the score of the path visited during the backup process and the number of visits. After the time for the update has elapsed, the most visited stone position in the MCTS tree is finally selected.

The process of updating the MCTS tree described above is, fortunately, one that proceeds via a distributed operation.

In its last development, AlphaGo performed MCTS by distributing computations across multiple CPUs and GPUs instead of performing calculations on a single computer, thus becoming a form of artificial intelligence with such a "godlike" degree of searching ability that it has proven outside the understanding of even professional Go players.

## 4.2  IBM Watson

Watson was named after IBM's founder and first CEO, industrialist Thomas J. Watson.

Watson is a question-answering computer system capable of answering questions posed in natural language, developed for IBM's DeepQA project by a research team led by principal investigator, David Ferrucci. The computer system was initially developed to answer questions on the quiz show *Jeopardy!* and was featured on its Februrary 14, 2011 episode. The first machine to win in a battle with humans was not AlphaGo. IBM's Deep Blue beat the 1996 World Chess Champion Garry Kimovich Kasparov. AI Watson, the upgraded version of Deep

Blue, won the *Jeopardy!* final championship after beating the American quiz show's two most successful champions in 2011.

In 2004, over dinner with coworkers, IBM Research manager Charles Lickel noticed that the restaurant they were in had fallen silent. He soon discovered the cause of this evening hiatus: Ken Jennings, who was then in the middle of his successful 74-game run on Jeopardy!. Nearly the entire restaurant had piled toward the televisions, mid-meal, to watch the phenomenon. Intrigued by the quiz show as a possible challenge for IBM, Lickel passed the idea on, and in 2005, IBM Research executive Paul Horn backed Lickel up, pushing for someone in his department to take up the challenge of playing Jeopardy! with an IBM system. Watson's actual development started in 2005 and proved to be a more difficult process than the chess iteration because it had to analyze the language in real time. Most importantly, it was the first artificial intelligence system to beat humans. Figure 4-5 shows the IBM "Man vs. Machine" Challenge.

IBM is expanding Watson into cognitive computing that combines artificial intelligence with machine learning to enable inference and learning. The fusion of big data and AI makes Watson a technology capable of analyzing huge amounts of data on its own and supporting human decision-making by accumulating knowledge while learning and evolving through human language. IBM Watson applies advanced natural language processing, information acquisition, knowledge representation, automated reasoning, and machine learning technologies to answer questions in unspecified areas.

**Figure 4-5**  Jeopardy! IBM "Man vs. Machine" challenge

(Source: http://stephsureads.blogspot.com/2011/02/jeopardy-ibm-challenge-and-what-it.html)

In addition, through machine learning, IBM Watson continues to learn for itself, developing its expertise. Watson is now evolving while interacting with human language through speech recognition, image recognition and visualization techniques.

Watson and AlphaGo are both deep learning technologies that play a prominent role in the field of health and games, respectively AlphaGo's core technology is a deep learning technology used to cluster or classify objects and data. AlphaGo uses deep learning to implement a value network, which calculates the odds of winning on the board, and a policy network, which calculates scores by location on the board.

Google DeepMind used TensorFlow, an in-depth learning system, to create AlphaGo. TensorFlow is an open source library for machine learning and deep learning.

This technology allows Google to adapt its general-purpose algorithm, AlphaGo, to a variety of fields. TensorFlow is the big data platform we all know.

Table 4-2 shows a comparison of IBM Watson Vs Google AlphaGo

Watson's core role is to build knowledge in specialized fields, such as medical terminology, and to generate optimal data to support human decision-making. When medical staff enter clinical information, Watson gives advice regarding a patient's condition and possible course of treatment. Based on data from millions of diagnostics, patient records, medical books, etc., Watson makes judgments by itself and informs users about the most effective course of treatment. Watson is helping to treat patients by reducing the time taken to interpret patient information and collect information from medical literature to just a few minutes.

**Table 4-2**   IBM Watson Vs Google AlphaGo

| IBM Watson | VS | AlphaGo |
|---|---|---|
| IBM | Development company | Google DeepMind |
| David Ferrucci | Developer | Demis Hassabis |
| Self-learning through machine learning | Learning method | Intensive learning using deep neural network technology (Value network + Policy network) |
| Won the Jeopardy quiz game in 2011 | Competitive Record | Won a Go game against Lee Sedol in 2016 |
| NLP (Cognitive computing) | Characteristic | Decision-making system |
| Cloud | System Used | TensorFlow |
| Medical, Finance | Main Application Field | Games, Health |

Watson's technical skills are high enough to analyze unstructured data such as pictures, videos, MRI data, and patient movements that have previously been difficult to analyze.

### – How does Watson work?

Following expert guidance, Watson collects the information necessary to develop corpus-based knowledge. Therefore, it can be said that the tasks of amassing such a corpus is to load relevant literature into Watson. In order to build a corpus, human intervention is required to look at the amassed information and to discard information that is out of date, poorly evaluated, or less relevant to the field.

This is called content curation. Content curation must be carried out effectively for Watson to learn properly. In order to learn how to interpret a great deal of information, expert training is required. To gain the ability to learn optimal response and search patterns, Watson becomes the expert's partner, and trains itself through machine learning. Machine learning is the application of AI to enable human-like reasoning, having started with Douglas Lenat in 1984 as CYC, an AI project which sought to create a database of comprehensive ontology and commonsense.

For example, take the statements "every tree is a plant", and "plants eventually die".

When asked whether the tree dies or not, the Inference Engine can draw definite conclusions and answer correctly. The expert uploads the training data to Watson, in question / answer format, which is used as verification data.

Figure 4-6 shows the Knowledge-Based Agent in Artificial Intelligence.

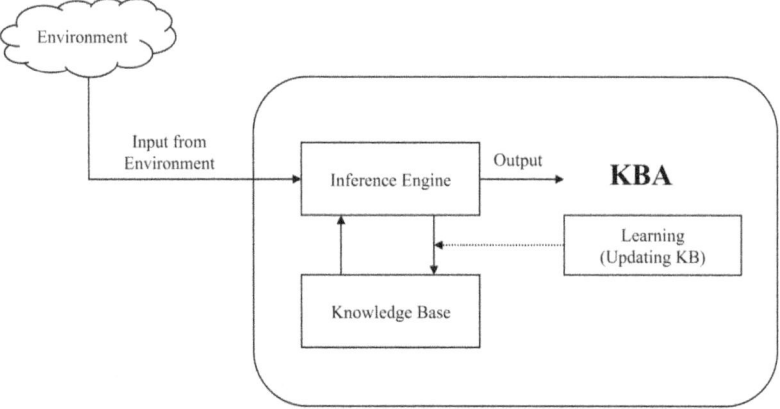

**Figure 4-6**   Knowledge-Based Agent in artificial intelligence

Experts regularly review the interaction between users and Watson and give feedback to the system to improve Watson's ability to interpret the information. These rules are human additions. It is unfeasible to make rules that encompass the entirety of the world's disparate knowledge. However, in Watson's case this was done automatically through natural language processing. After all, Watson is a system that continually updates itself when new information is released and is always learning new things by processing the language in which that knowledge is written.

## 4.3 Practice Questions

Q1. What is AlphaGo?

Q2. Describe the Monte Carlo Tree Search (MCTS) which is implemented as the AlphaGo algorithm.

Q3. Describe the AlphaGo Pipeline Structure.

Q4. Describe Watson, an artificial intelligence system implemented by IBM.

# Chapter 5

# Microsoft Azure Machine Learning Studio

## 5.1 Introduction of Microsoft Azure Machine Learning Studio

Azure is a cloud computing platform from Microsoft that has been in service since 2010. Following PaaS in 2011, IaaS service was launched in 2013. Of the more than 600 services that the Azure platform provides, this book will focus on only one of them, Azure Machine Learning Studio. There are many open source libraries and tools for machine learning, so why use Azure Machine Learning Studio? The answer lies in the fact that Azure Learning Studio solves the following problems.

(1) Lack of flexibility in computing resources for learning
(2) Difficulties in configuring GPU-based environments for machine learning
(3) Difficulty installing and setting up the tools necessary for learning
(4) Difficulty in experiment recording and versioning

Figure 5-1 shows the workflow of Azure Machine Learning. Users can leverage the Azure platform to enable data collection and management in the cloud and use Machine Learning Studio to create predictive models and build web services easily. Unlike the existing cloud platform and machine learning libraries and tools, it provides an easy-to-access GUI environment with user convenience in mind.

Figure 5-2 shows that Azure ML Studio can easily create blocks in the form of drag-drop, unlike traditional machine learning tools and libraries. In addition, scripts written in R and Python languages can be inserted and used in blocks, and

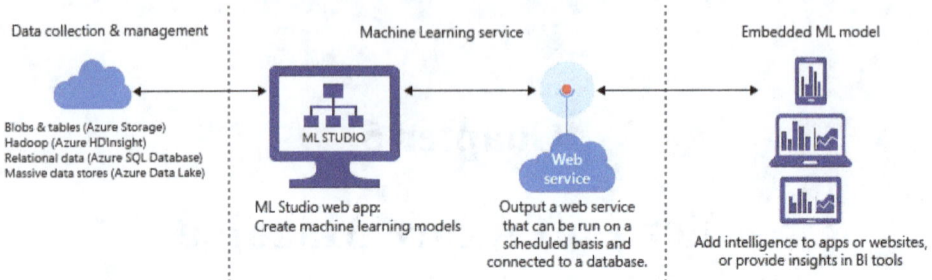

**Figure 5-1**   Azure Machine Learning Basic workflow

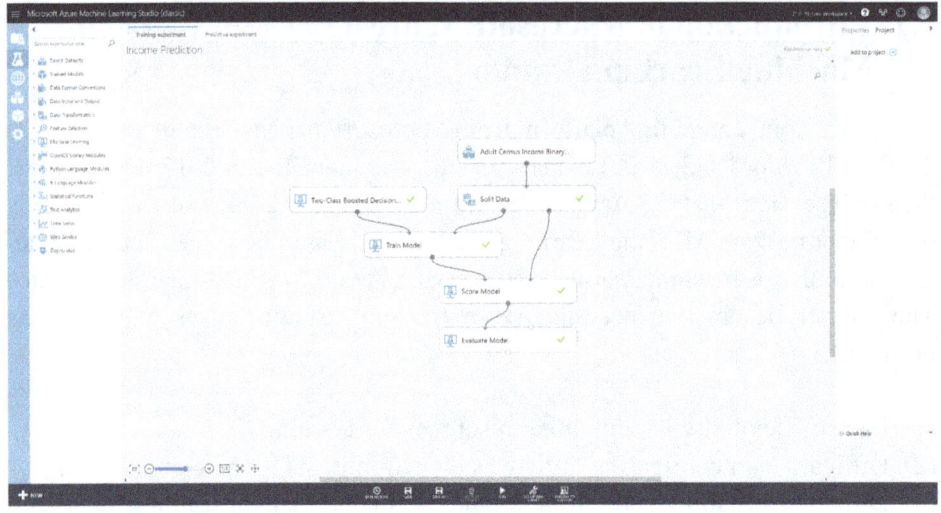

**Figure 5-2**   Canvas of Azure Machine Learning Studio Experiment

the results can be visualized. Thanks to its easy structure, users of all skill levels can easily create and distribute predictive models.

Figure 5-3 shows the Azure AI Gallery where users can try out projects developed by other users or import them into their own projects for modification. In addition, users can post projects shared in the Gallery on social media, such as Facebook or Twitter, to request comments.

Figure 5-4 shows the overall capabilities of Azure Machine Learning. Azure Machine Learning supports data input, output and visualization, and prepares standard machine learning algorithms frequently used by data scientists. Users can develop and deploy the following steps.

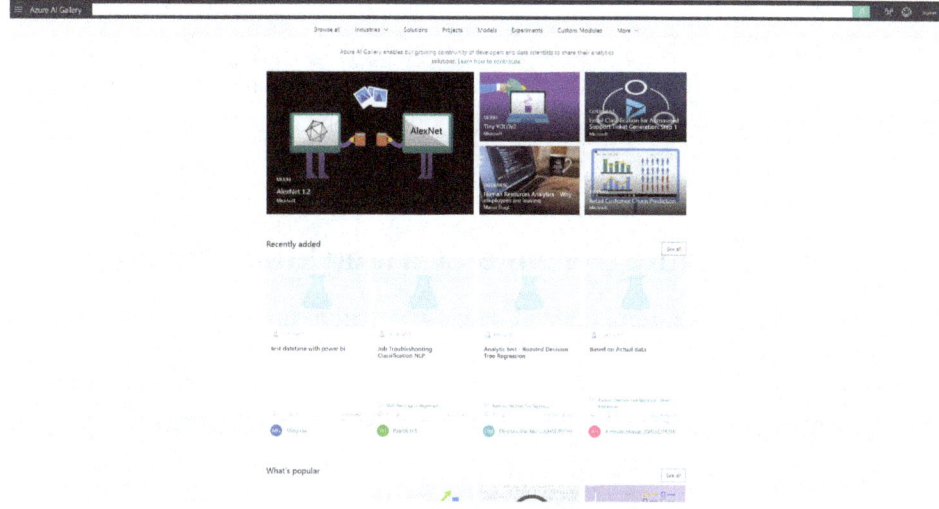

**Figure 5-3**   Azure AI Gallery

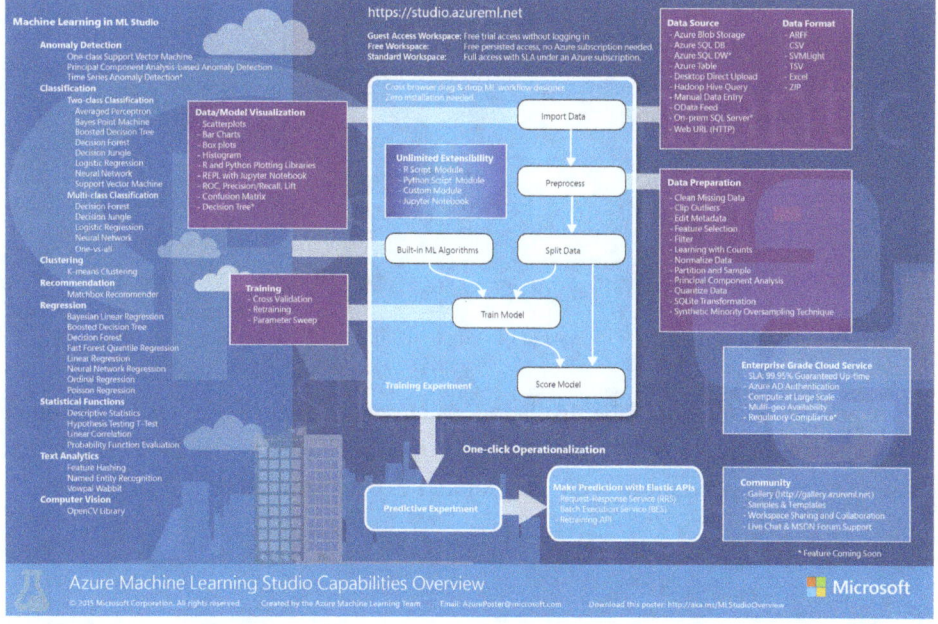

**Figure 5-4**   Diagram of Azure Machine Learning Studio

(1) Import data: Users can take data for training models and import it into the Azure cloud for versatile use.

(2) Data preprocessing: Users can preprocess data for situations in which there is missing data for training models.

(3) Feature selection: To train a model, users can extract features according to the algorithm.
(4) Evaluation: After training, the model is evaluated and checked for errors.
(5) Web service deployment: The trained model can be distributed through a web service.

## 5.2  Microsoft Azure Machine Learning Studio Sign-up

Go to https://studio.azureml.net/and click the 'Sign up here' link, boxed in Figure 5-5.

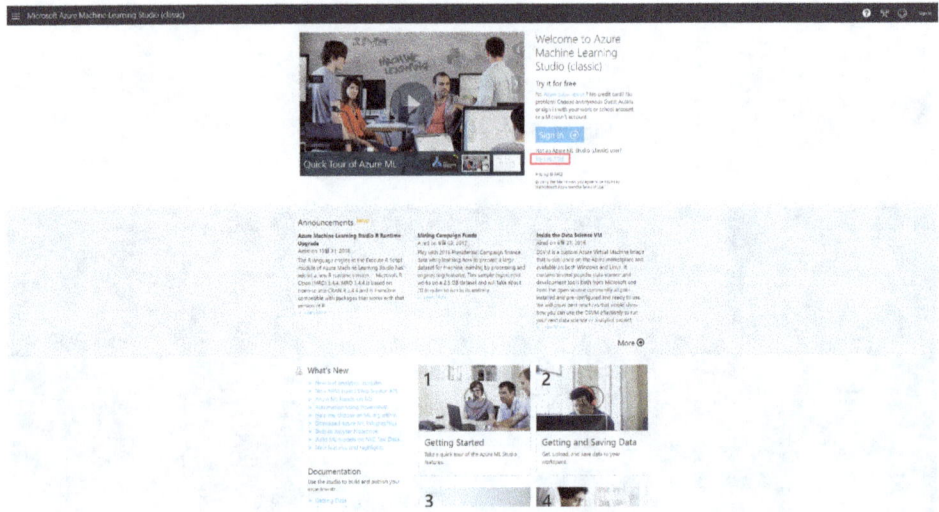

**Figure 5-5**   Homepage of Azure Machine Learning Studio

A pop-up window will be displayed, as shown in Figure 5-6. Users can enter as a guest for 8 hours free of charge, but it is recommended to use Free Workspace with a Microsoft account to store project and experiment data.

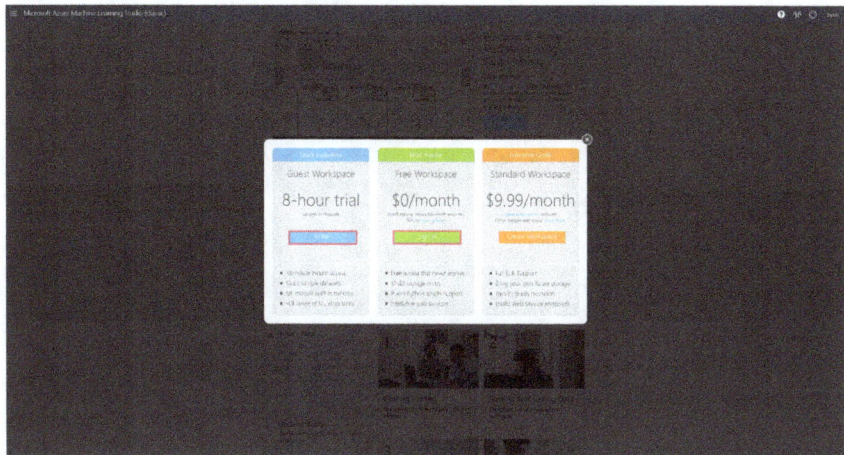

**Figure 5-6**   Azure Machine Learning Studio user selection pop-up

If you log in successfully, you will see the workspace shown in Figure 5-7.

**Figure 5-7**   Azure Machine Learning Studio workspace

## 5.3  Introduction of Microsoft Azure Machine Learning Studio Function

Click the "+ NEW" button in Figure 5-8 to display the Workspace features

**Figure 5-8**    "+ NEW" button in Azure Machine Learning Studio workspace

Click the "+NEW" button to see the "DATASET", "MODULE", "PROJECT", "EXPERIMENT" and "NOTEBOOK" functions as shown in Figure 5-9. The 'DATASET' function allows users to upload experiment data files to the Azure cloud for use in their personal account. Users can also modularize Python scripts by inserting a compressed file (.zip) containing modules written in the Python language.

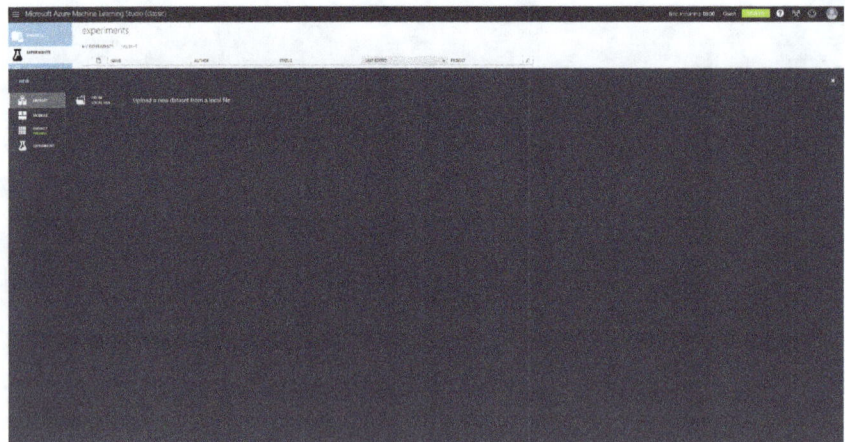

**Figure 5-9**    DATASET function

The "MODULE" function in Figure 5-10 can be used to modularize R scripts by inserting a compressed file (.zip) containing scripts written in R languages. Users can also import and use module examples written in various R languages that Microsoft has published in the Gallery.

**Figure 5-10** MODULE function

**Figure 5-11** PROJECT function

Using the "PROJECT" function in Figure 5-11, users can conveniently manage the datasets, experiments, modules, notebooks, trained models, and 'webservices' as a group.

The "EXPERIMENT" function shown in Figure 5-12 is a place where users can make models and deploy 'webservices' by using the inserted dataset and various functions. Users can also import and use the examples that Microsoft publishes in the Gallery.

Table 5-1 lists the functions of the Microsoft Azure Machine Learning Studio Experiment. Find the appropriate function and use it for your purpose.

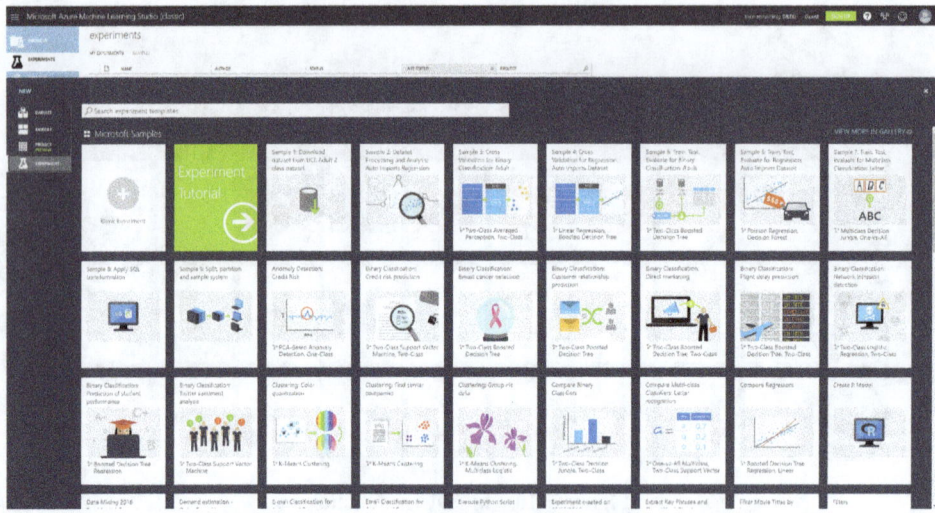

**Figure 5-12**   EXPERIMENT function

**Table 5-1**   Azure Machine Learning Experiment functions

| Function | Function explanation |
| --- | --- |
| Saved Datasets | Repository containing data sets and sample data sets saved using DATASET function |
| Trained Models | Repository containing trained models |
| Data Format Conversions | The function of converting the format of the data (.arff, .csv etc.) |
| Data Input and Output | Data input, output, and import of training models |
| Data Transformation | Preprocessing (Split, Filter, Join, etc.) to train the data |
| Feature Selection | To extract a specified number of features based on statistics or machine learning algorithms |
| Machine Learning | Contains the algorithms used to create the model. There are Machine Learning algorithms, training, and evaluation models. |
| OpenCV Library Modules | There is OpenCV Library function which is an image processing library. |
| Python Language Modules | It is possible to run Python scripts. |
| R Language Modules | It is possible to run R scripts. |
| Statistical Functions | It contains statistical functions such as descriptive statistics, statistical tests, correlation matrix calculation, and recode. |
| Text Analytics | It is possible to use Text Analytics. |
| Time Series | It is possible to use Time Series Anomaly detection. |
| Web Service | Input/output data management for web services is possible. |
| Deprecated | Collection of functions to disappear. |

The algorithms in Table 5-2 can be conveniently imported into blocks in the navigation palette. If users cannot find a desired algorithm, they can implement it using Python and R scripts, and then place blocks containing the implemented algorithm in the experiment.

**Table 5-2**    Azure Machine Learning Algorithm (2018.1)

| Algorithms Classification | Algorithms |
| --- | --- |
| Anomaly Detection | One-Class Support Vector Machine |
| | PCA-Based Anomaly Detection |
| Classification | Multiclass Decision Forest |
| | Multiclass Decision Jungle |
| | Multiclass Logistic Regression |
| | Multiclass Neural Network |
| | One-vs-All Multiclass |
| | Two-Class Averaged Perceptron |
| | Two-Class Bayes Point Machine |
| | Two-Class Boosted Decision Tree |
| | Two-Class Decision Forest |
| | Two-Class Decision Jungle |
| | Two-Class Locally Deep Support Vector Machine |
| | Two-Class Logistic Regression |
| | Two-Class Neural Network |
| | Two-Class Support Vector Machine |
| Clustering | K-Means Clustering |
| Regression | Bayesian Linear Regression |
| | Boosted Decision Tree Regression |
| | Decision Forest Regression |
| | Fast Forest Quantile Regression |
| | Linear Regression |
| | Neural Network Regression |
| | Ordinal Regression |
| | Poisson Regression |

## 5.4  Practice Questions

Q1. Explain the advantages of Azure Machine Learning Studio.


Q2. Explain the algorithms supported by Azure Machine Learning Studio.

# Chapter 6

# Create Prediction Model using Microsoft Azure Machine Learning Studio

## 6.1 Microsoft Azure Machine Learning Studio Experiment Configuration and Functions

Before moving onto the tutorial, let's take a look at the screen configured in the experiment. Check the function description by the number shown in Figure 6-1.

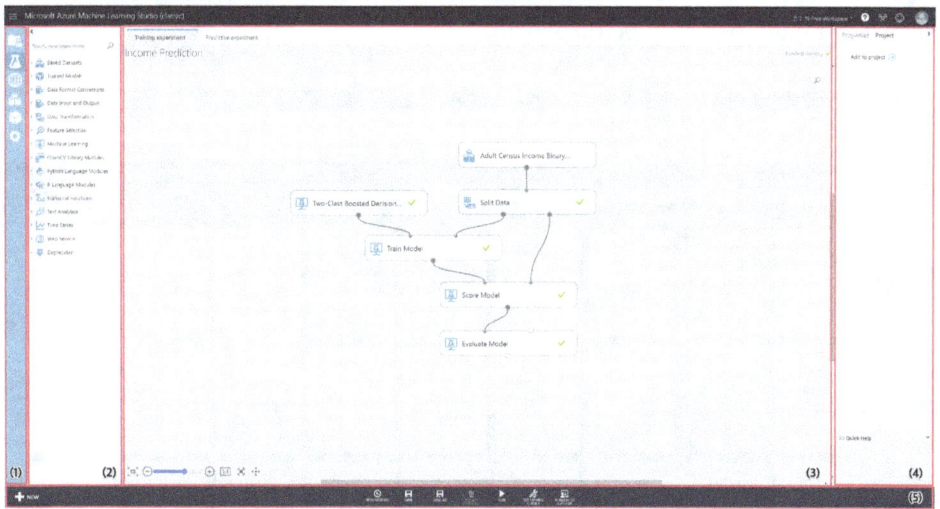

**Figure 6-1** Azure Machine Learning Studio Experiment configuration

(1) Category

Sidebar for managing datasets, experiments, etc.

(2) Palette

Area containing the main functions of the project.

(3) Experiment Canvas

The drag-and-drop workflow area, in which models can be created and real analysis performed.

(4) Properties

Displays the properties of each module. Each module can have different properties.

(5) Option Menu

Users can save, execute and publish web services, galleries, etc. The "+NEW" button can be selected to insert datasets, import modules, etc.

## 6.2  Microsoft Azure Machine Learning Studio Experiment Tutorial

Microsoft provides a tutorial for Azure Machine Learning Studio. If users click the "+ NEW" button and enter the "EXPERIMENT" menu, "Experiment Tutorial", as shown in Figure 6-2, will be displayed.

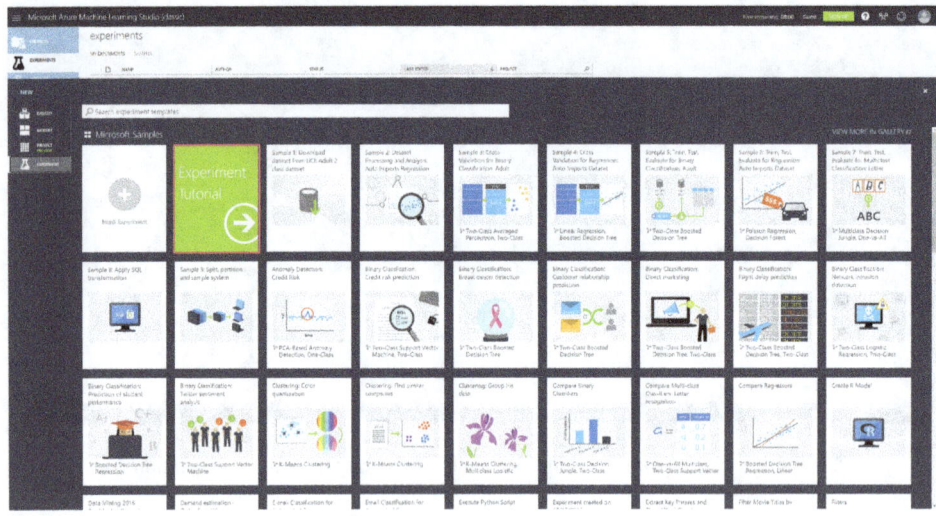

**Figure 6-2**   Experiment Tutorial

Users can start the tutorial by clicking the "Experiment Tutorial" button. As blocks are dragged and dropped automatically, explanations are provided.

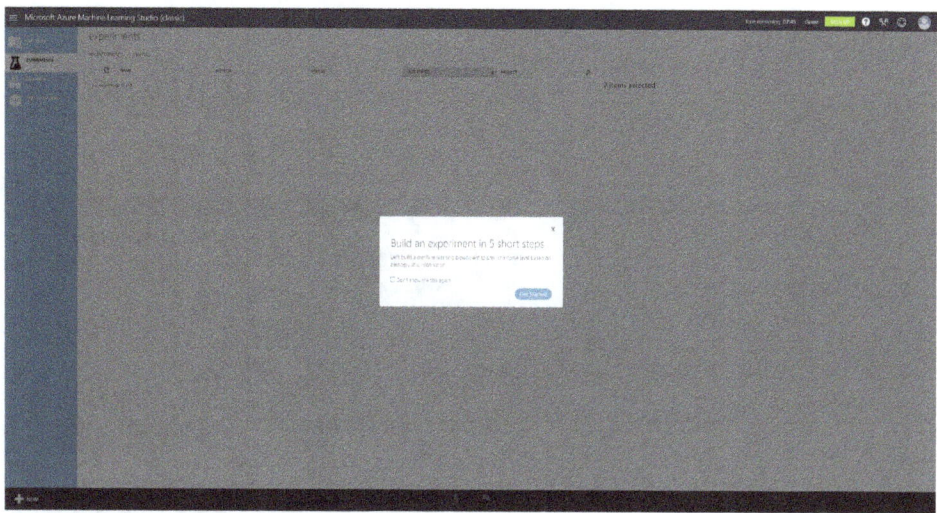

**Figure 6-3** Five-step tutorial

Click the "Get Started" button in Figure 6-3 to start the 5-step tutorial.

Figure 6-4 shows the beginning of the tutorial in which the mouse automatically moves to create an empty experiment. Without an experiment, users cannot create prediction models using Azure Machine Learning Studio.

**Figure 6-4** Remote operation tutorial

In Figure 6-5 the experiment name has been changed to "Income Prediction", and the sample data is being imported to make the model. Azure Machine Learning Studio provides sample data for the tutorial. Users proceed by clicking the "Show Me" button.

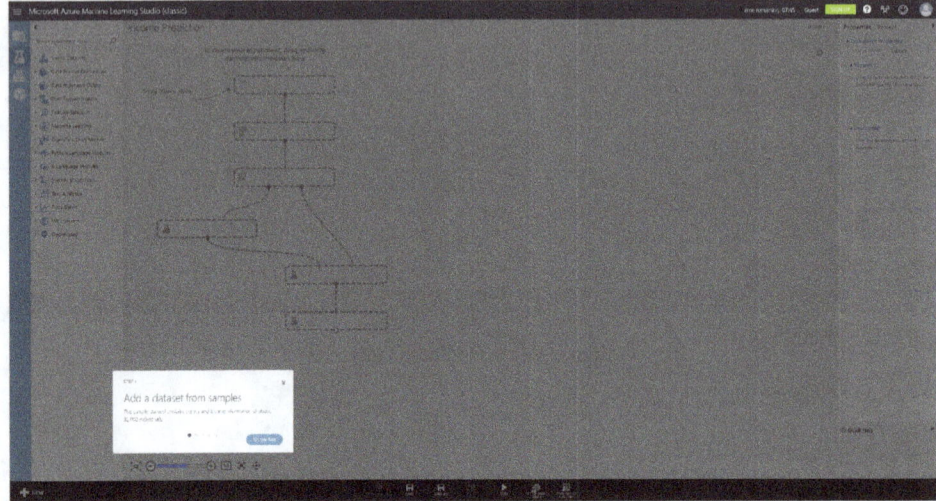

**Figure 6-5**  Import sample data

In Figure 6-6 "Income" has been retrieved from the palette and the "Adult Census Income Binary Classification Dataset" a sample Saved Dataset, has been imported. Users will learn how to insert the data they want later.

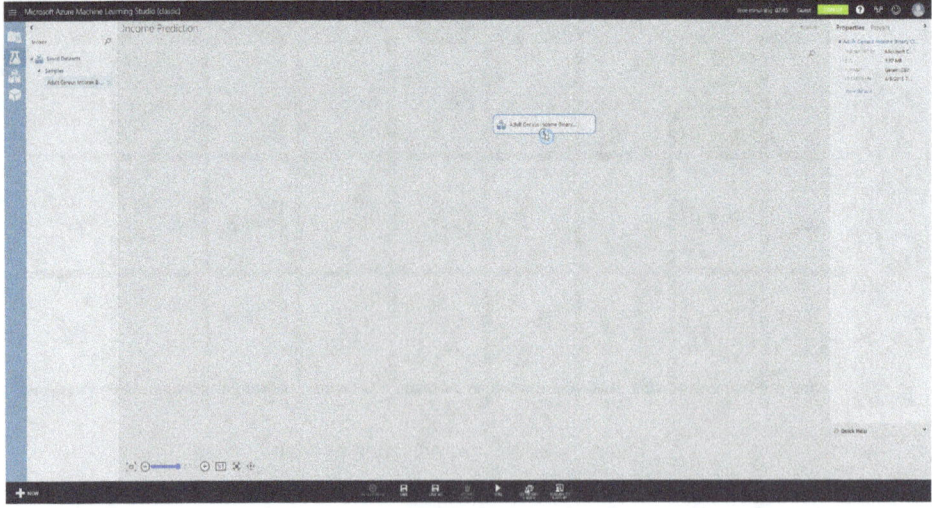

**Figure 6-6**  Provide sample data in the palette

As shown in Figure 6-7 right-clicking on the imported data set block and clicking the "Visualize" button will display a visualized data set. Users can also visualize their data set by simply clicking on it.

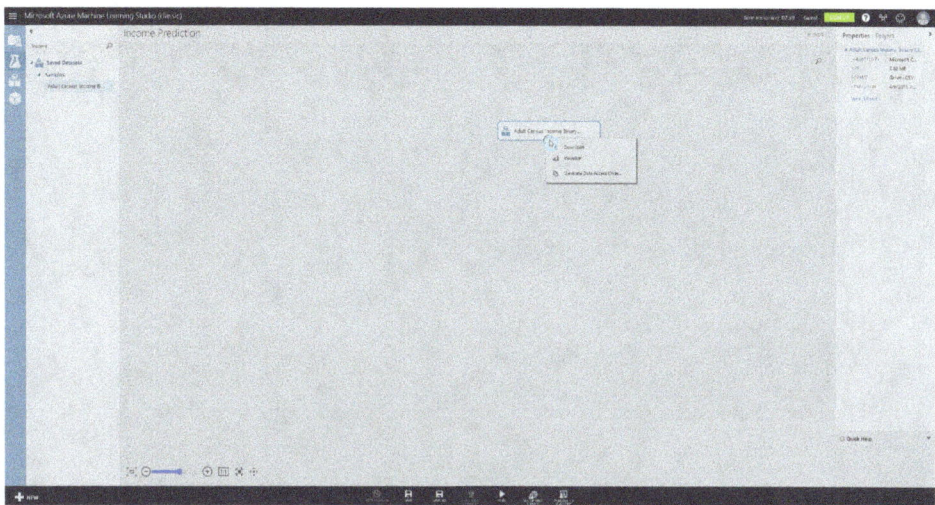

**Figure 6-7** Data visualization

Figure 6-8 shows data rows and columns through data visualization, displaying data attributes in detail.

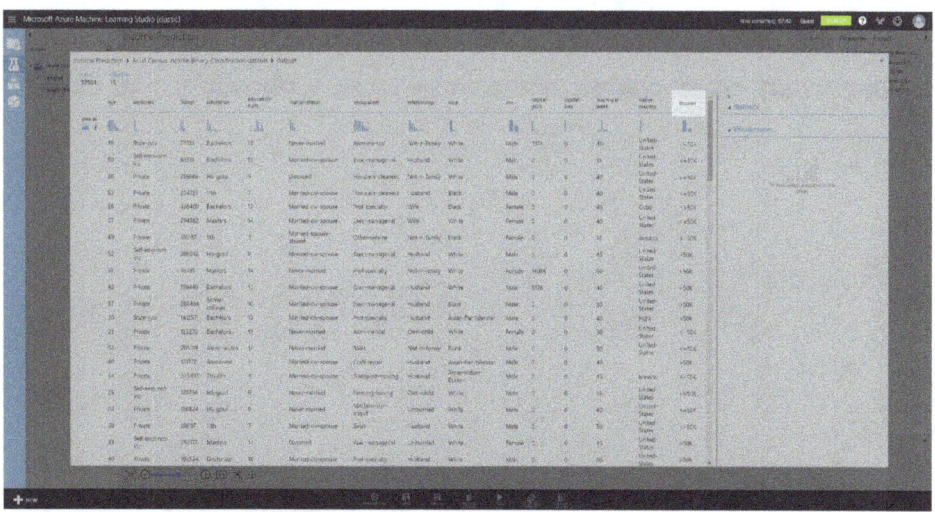

**Figure 6-8** Attribute detail via data visualization

The screen in Figure 6-9 suggests splitting the data set to prevent overfitting. When training the model, it is crucial to divide the training and test data into a data set to avoid overfitting. Usually, there is a tendency to divide the training and test data into a 7: 3 ratios.

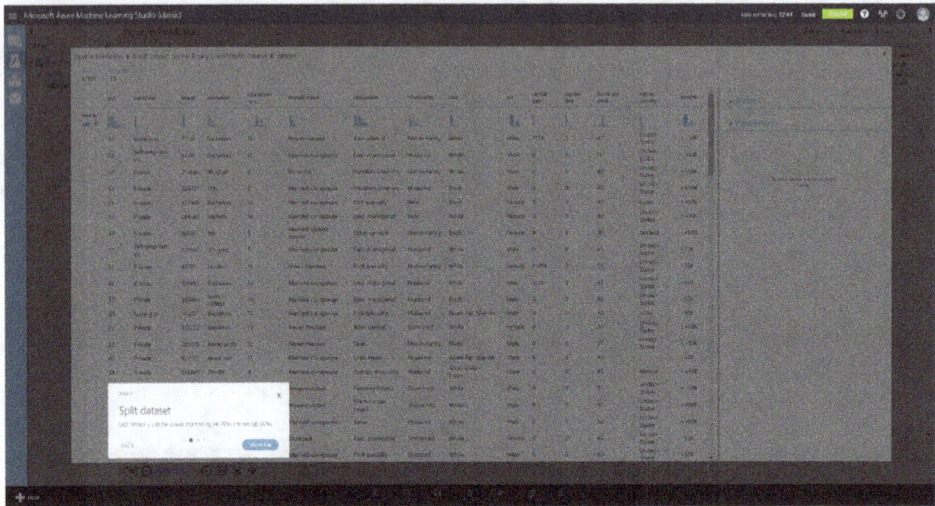

**Figure 6-9**   Dataset split

Figure 6-10 displays searching for a split using the search bar, placing it on the canvas, and setting the split ratio of the training data and the random seed

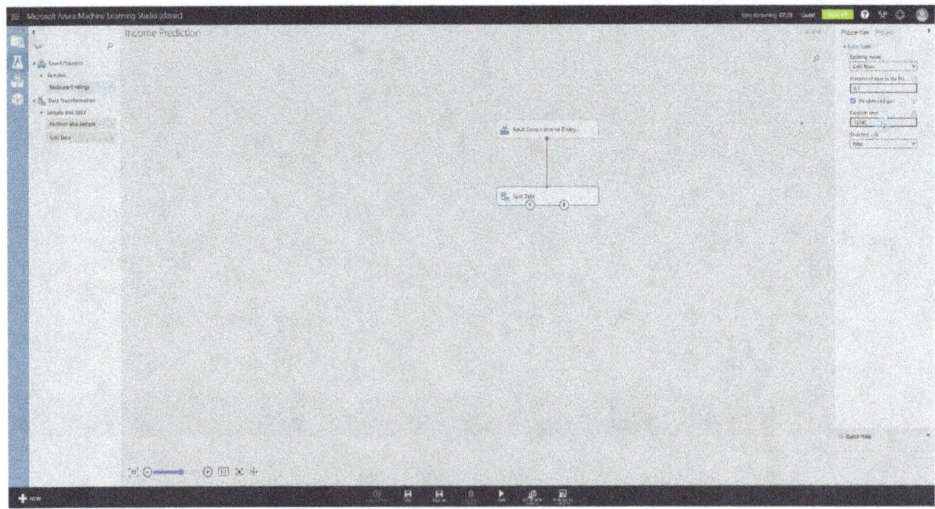

**Figure 6-10**   Random seed for dataset split

parameter in the "Split" property window. Enter a training data rate of 0.7 and enter a random seed value. The random seed is a parameter value for selecting randomly because the computer itself is logically organized.

The screen in Figure 6-11 shows that the dataset split is complete and suggests an algorithm to train the model. This tutorial applies the Two-Class Boosted Decision Tree algorithm. The word to watch is "Two-Class". Why should you pay close attention to the word "Two-Class"? Let's take a look slowly as we go through the tutorial.

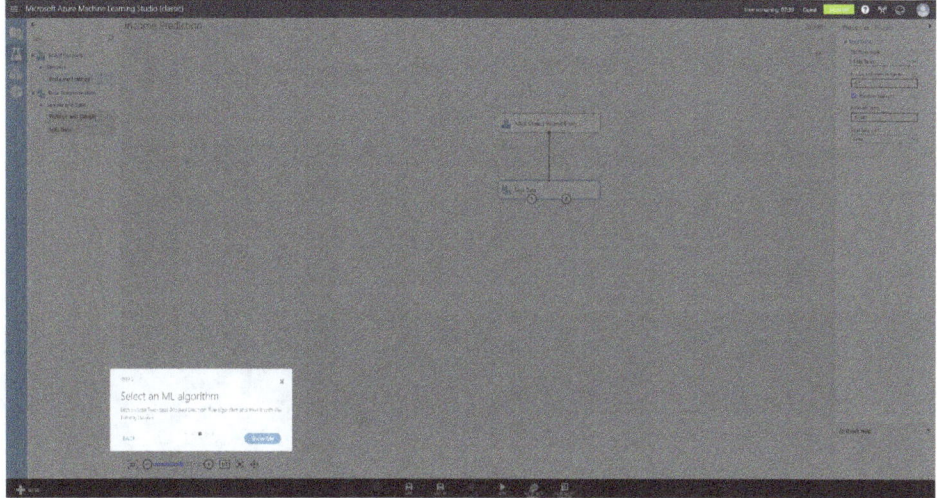

**Figure 6-11**   Algorithm Selection for Prediction Model training

Figure 6-12 shows the Two-Class Boosted Decision Tree algorithm placed on the canvas on the palette, and the training model placed on the palette in order to train the data. Close readers may have noticed the right exclamation point in the "Train Model" block. An error has occurred because the Label Column was not defined when applying the Two-Class Boosted Decision Tree. Let's set the Label Column by clicking "Launch column selector" in the Properties window.

As Figure 6-13 shows, by clicking "Launch column selector", the Label Column is set to "Income". Why was "Income" chosen among so many attribute values? The reason for this is that the property value of "Income" is composed of <= 50K, > 50K, and therefore, of the many attribute values, the predicted value is one of the two related to "Income" This may go some way in answering the earlier question of 'Why Two-Class?'

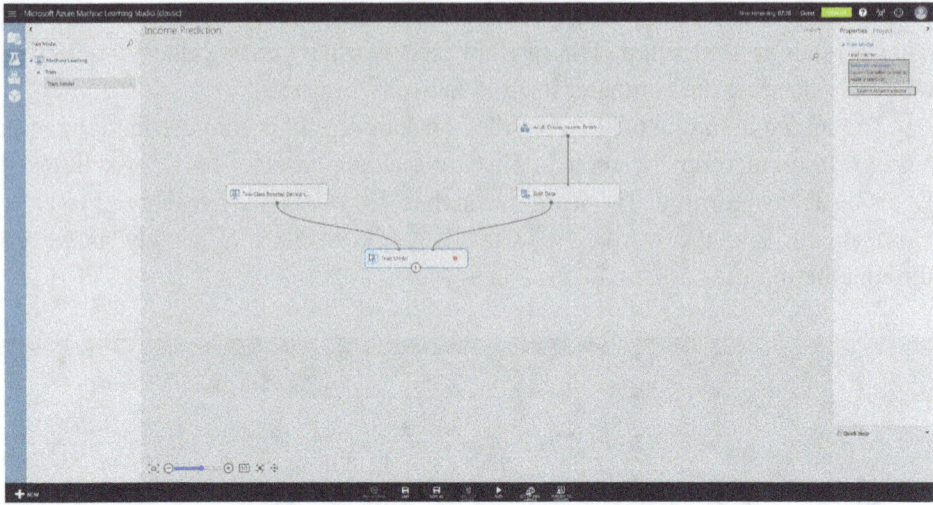

**Figure 6-12**   Training Model for Learning

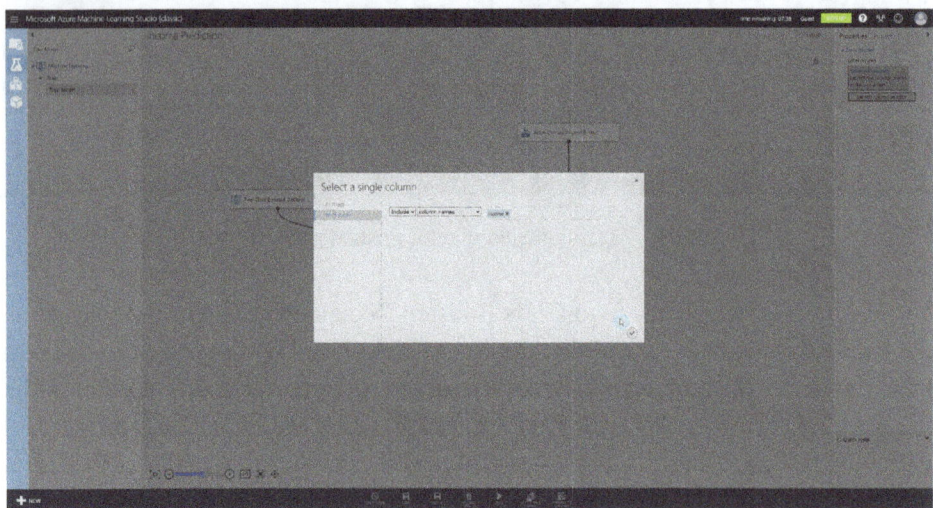

**Figure 6-13**   Prediction value selection

Figure 6-14 shows the error disappearing after selecting the "Income" column.

In Figure 6-15, the program suggests making a prediction using a training model and test data. Click the "Show Me" button to proceed.

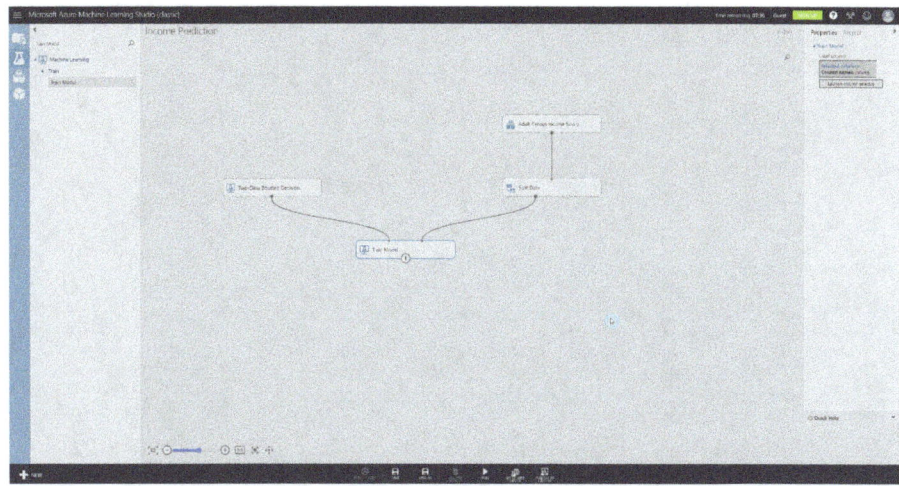

**Figure 6-14**   Error disappeared

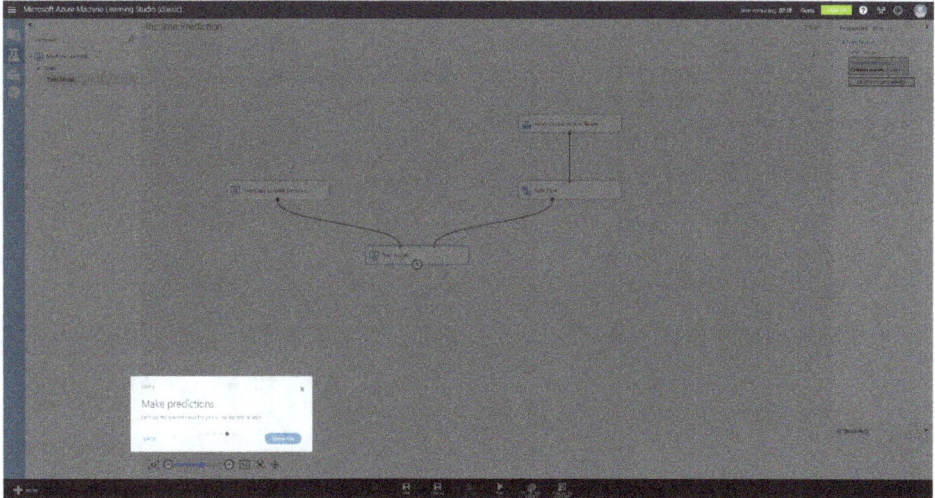

**Figure 6-15**   Propose prediction after training is complete

In Figure 6-16, the "Score Model" block is placed on the palette for prediction to take place, and the "Evaluate Model" block is placed and then connected to evaluate the predicted value.

In Figure 6-17 the program suggests running both training and prediction by clicking the "Run" button, found on the bottom bar, because the model for the prediction is complete. Click the "Show Me" button to proceed.

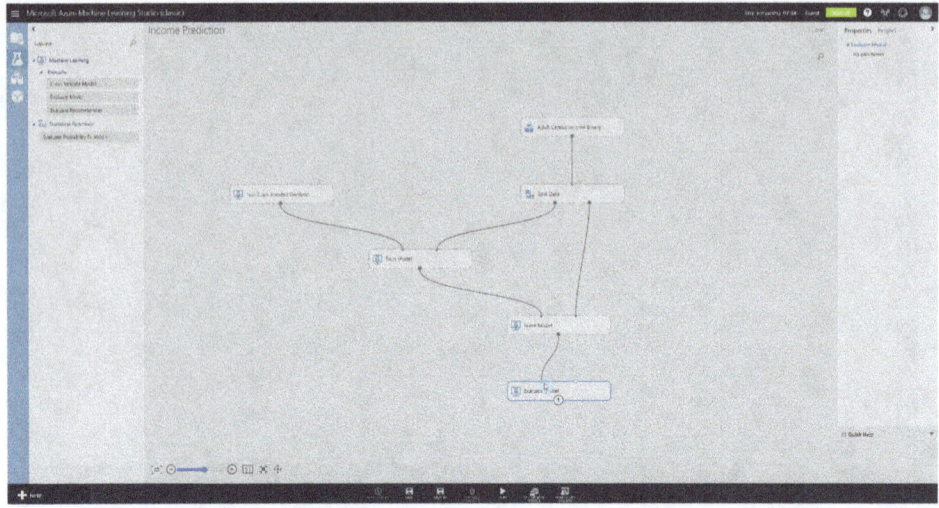

**Figure 6-16**   Arrangement for a prediction model

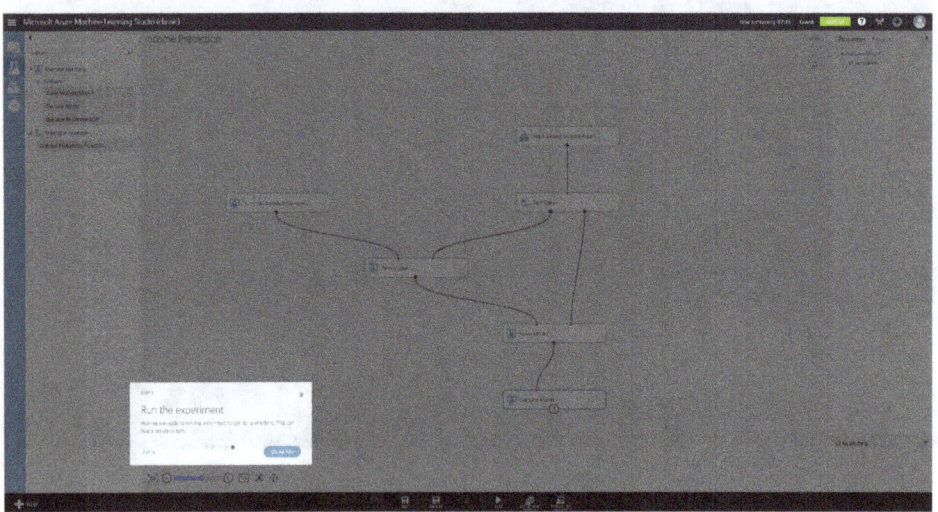

**Figure 6-17**   Suggest predicting by clicking 'Run'

Figure 6-18 shows the model training initiated by pressing the "Run' button". It is a tutorial, but a server in an Azure cloud center somewhere in the world is actually doing the operation.

Figure 6-19 shows a checkmark on all blocks except the data set, indicating that the model has been trained and the prediction has been carried out.

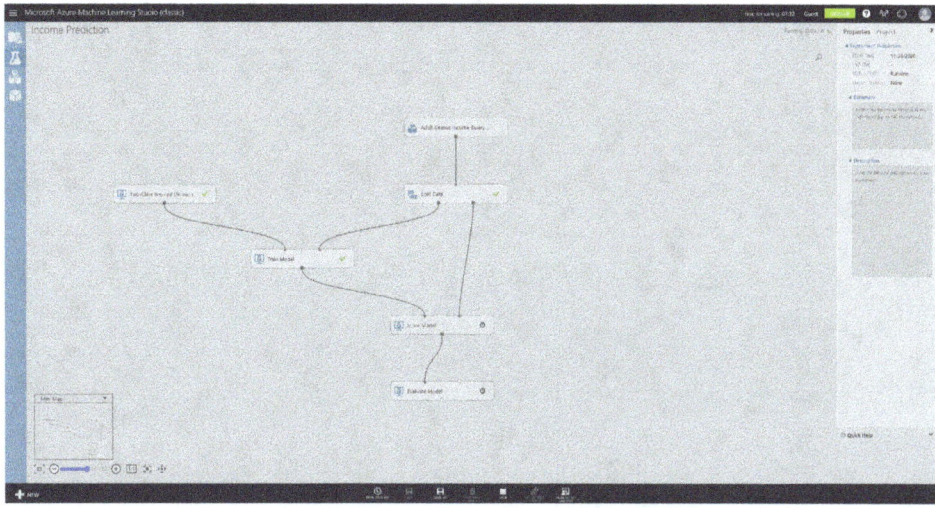

**Figure 6-18** Prediction model training

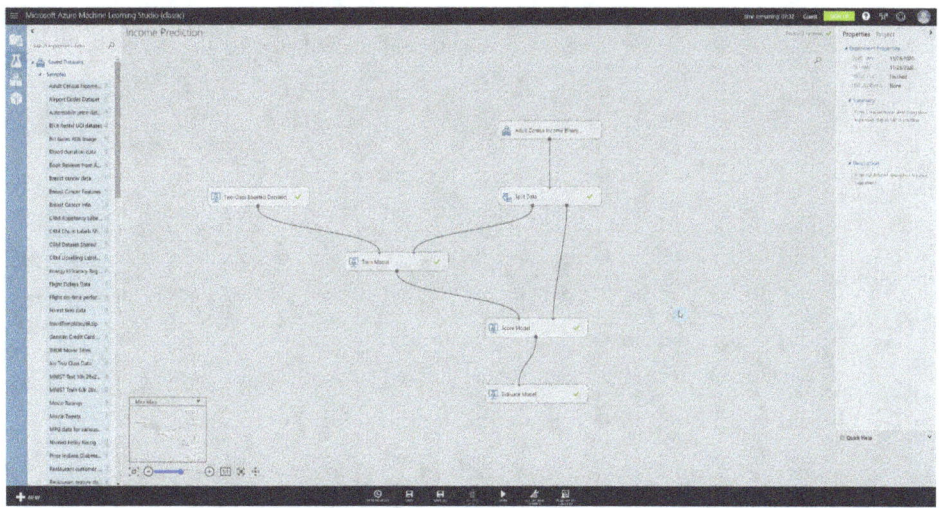

**Figure 6-19** Prediction Model Prediction complete

Figure 6-20 displays the process of viewing the Score Model status with the training of the model having been completed and prediction using the model having taken place. All models can be trained and predictions can be carried out via the "Visualize" item, which can be accessed with a right-click.

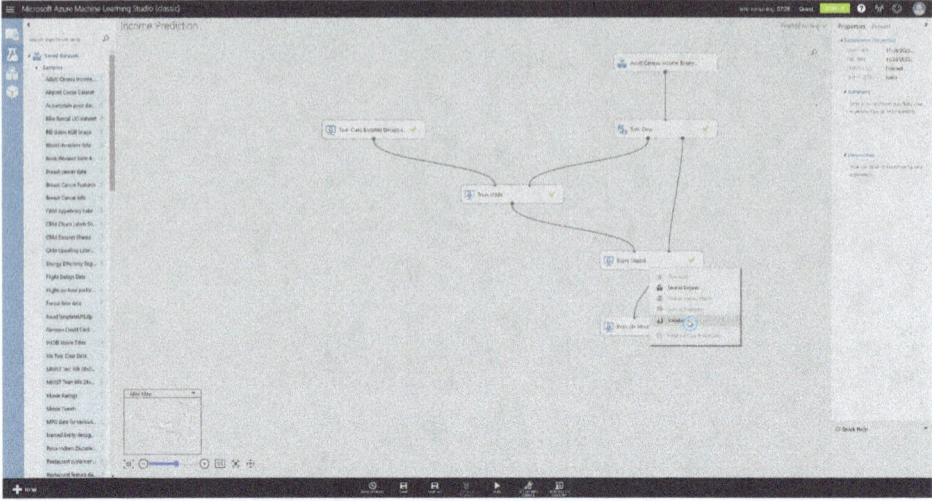

**Figure 6-20**　View predicted results

**Figure 6-21**　Check the predicted value

Figure 6-21 shows the process of checking Scored Labels and Scored Probabilities calculated by the Score Model. You can insert test data and view the predicted value and the probability of the predicted value through the Scored Model.

Figure 6-22 displays the performance of Score Model. You can right-click the "Evaluate Model" block to see the value that you evaluated through the "Visualize" item.

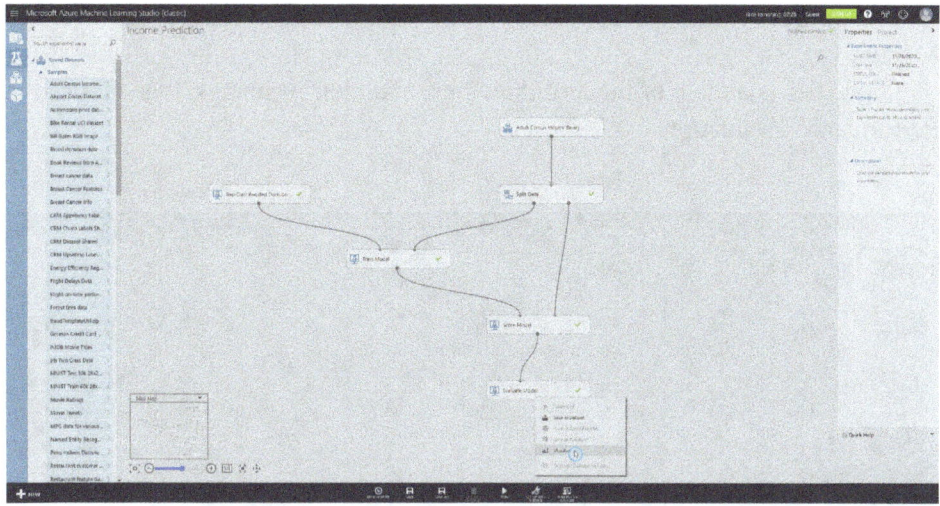

**Figure 6-22**　Verify the performance of the prediction model

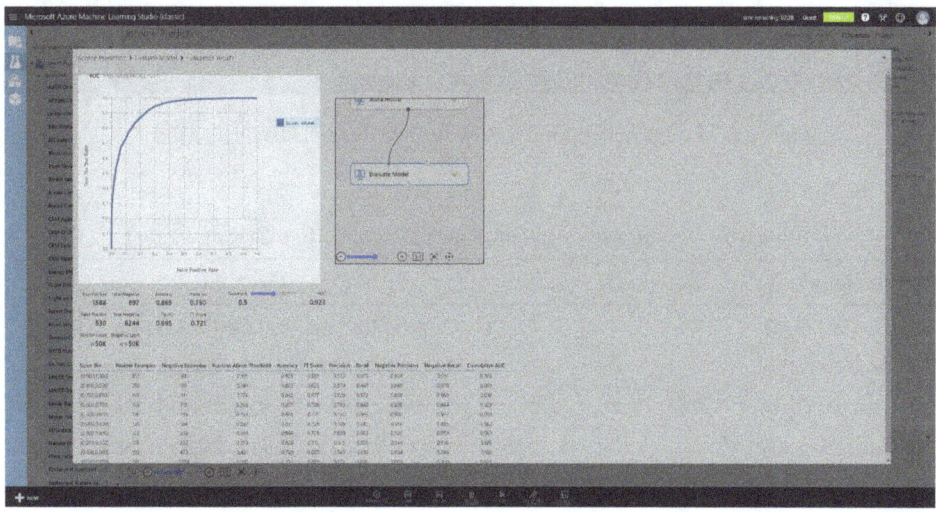

**Figure 6-23**　ROC Curve of Predictive Model through "Evaluate Model"

Figure 6-23 shows the receiver operating characteristic curve of the model. The ROC curve is a graph which shows the ratio of true positives to false positives. It is a visualization of the predictive performance of a binary classification model. The larger the area under the ROC, the better the performance. Of course, if the area of ROC is 1, you should suspect overfitting.

Figure 6-24 shows the evaluation result of model performance via 'Evaluate Model'.

Table 6-1 is a table of the terminologies, and their meanings, that are used in Performance Evaluation.

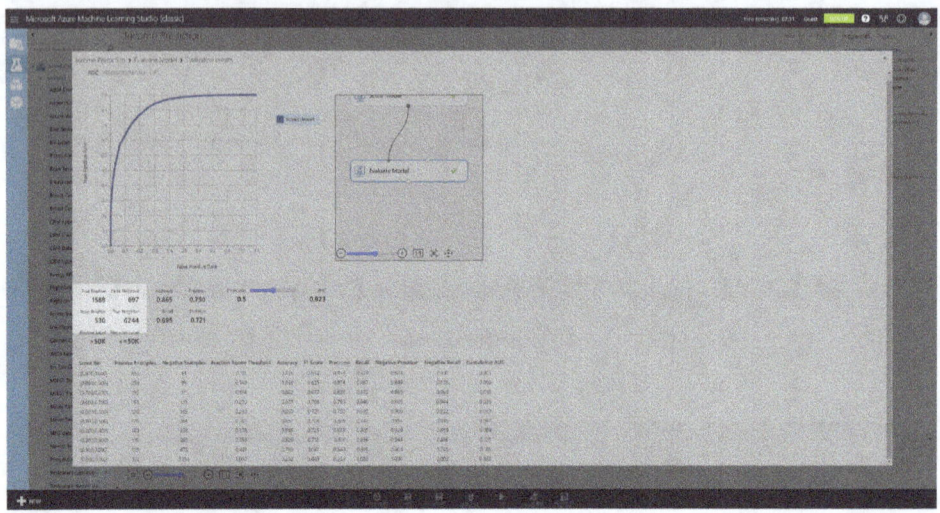

**Figure 6-24**    Prediction Model Performance through "Evaluate Model"

**Table 6-1**    Terminologies and meanings used in performance evaluation

| Terminology | Meaning |
|---|---|
| TP (True Positive) | If you predicted true and the data is true |
| TN (True Negative) | If you predicted true and the data is false |
| FP (False Positive) | If you predicted false and the data is true |
| FN (False Negative) | If you predicted false and the data is false |
| Accuracy — (TP+TN)/(TP+TN+FP+FN) | An indicator of whether the model finds the right prediction. |
| Precision — TP/(TP+FP) | The ratio of what the model classifies as true to what is actually true. |
| Recall — TP/(TP+TN) | The ratio of what is actually true to what the model predicts is true. |
| F1 Score | When data labels have an unbalanced structure, you can accurately evaluate the performance of the model and express the performance as a single number. |

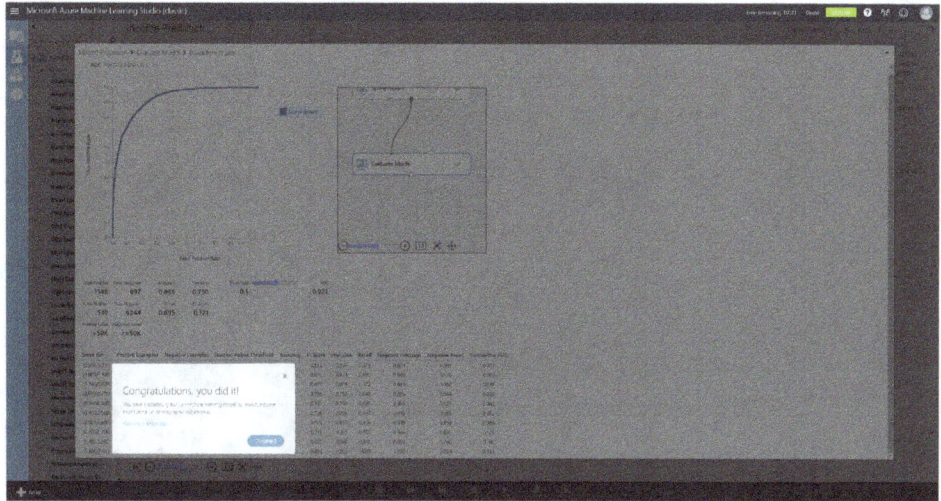

**Figure 6-25**   Tutorial complete

Figure 6-25 shows the completion of the prediction model creation tutorial. Although the tutorial lasts only about three minutes, it is enough time to learn the process of creating a prediction model. Once users have created the prediction model through the tutorial, they can make it available remotely by publishing it via the web service. Click the "Set up a web service" button to proceed. Take a look at the web service deployment tutorial in the next section.

## 6.3  Practice Making Microsoft Azure Machine Learning Studio Experiment Prediction Models

### 6.3.1  *Importing Data to Azure Cloud*

Azure Machine Learning Studio is cloud-based, so if you want to build a model based on your own data, you need to import this data into the cloud for model training. To do this, locate and click the "+ NEW" button which should be at the bottom. Then, click "From Local File", which should bring up the "Dataset" category, as shown in Figure 6-26.

Select and insert a local file, as shown in Figure 6-27.

**Figure 6-26**   Data import

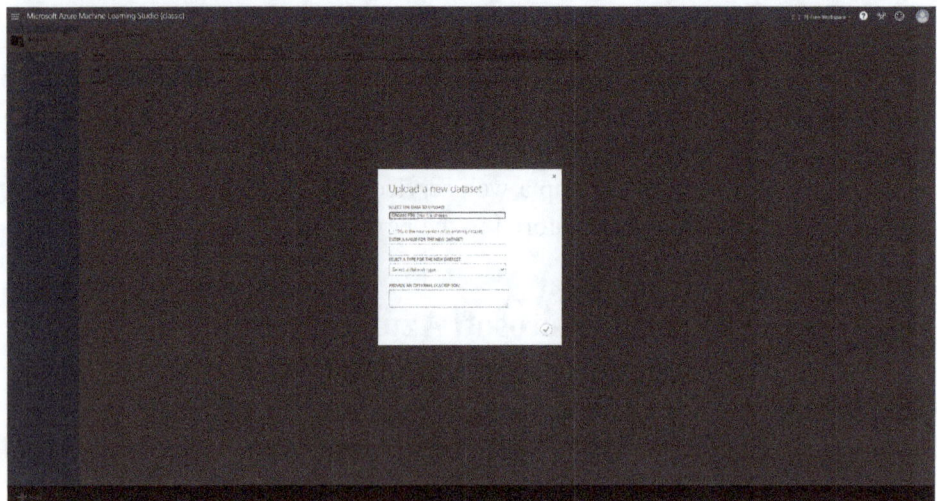

**Figure 6-27**   Data import popup window

The file type is automatically selected based on the file extension listed in Table 6-2. Extensions are automatically recognized and retrieved from the list below.

The imported data can be checked in the DATASETS menu shown in Figure 6-28.

When data import is complete, you can check the palette window, as shown in Figure 6-29. Of course, you can use the imported data freely in the experiment.

**Table 6-2**   Supported file extension

Generic CSV File with a header (.csv)

Generic CSV File with no header (. nh.csv)

Generic TSV File with a header (.tsv)

Generic TSV File with a header (.nh.tsv)

Plain Text (.text)

SvmLight File (.svmlight)

Attribute Relation File Format (.arff)

Zip File (.zip)

R Object or Workspace (.RData)

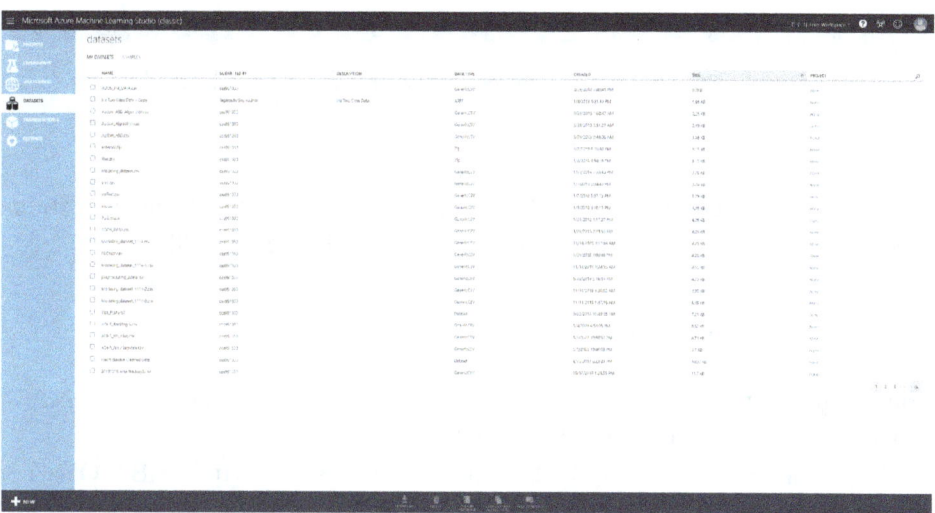

**Figure 6-28**   DATASETS Menu

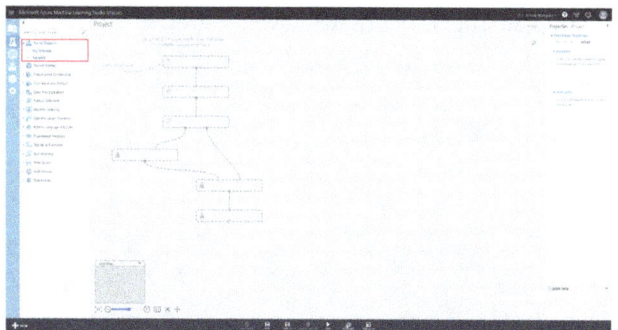

**Figure 6-29**   Data import status of palette

**Figure 6-30**    Sample data provided by Azure Machine Learning Studio

You can also check the sample data provided by Azure Machine Learning Studio by accessing the "Samples" menu, as shown in Figure 6-30.

### 6.3.2  *Visualize Data Set*

Prior to making a prediction model, proceed with the data in the link below. Download and import the data to proceed.

https://drive.google.com/file/d/1Ujp-SnVflA4USJbnUWqdiNzlB3SOWeY3/view?usp=sharing

Place the dataset on the canvas, as shown in Figure 6-31, right-click, and then click "Visualize".

As shown in Figure 6-32, you can check the number of rows and columns in the dataset, and you can check the distribution with the desired model.

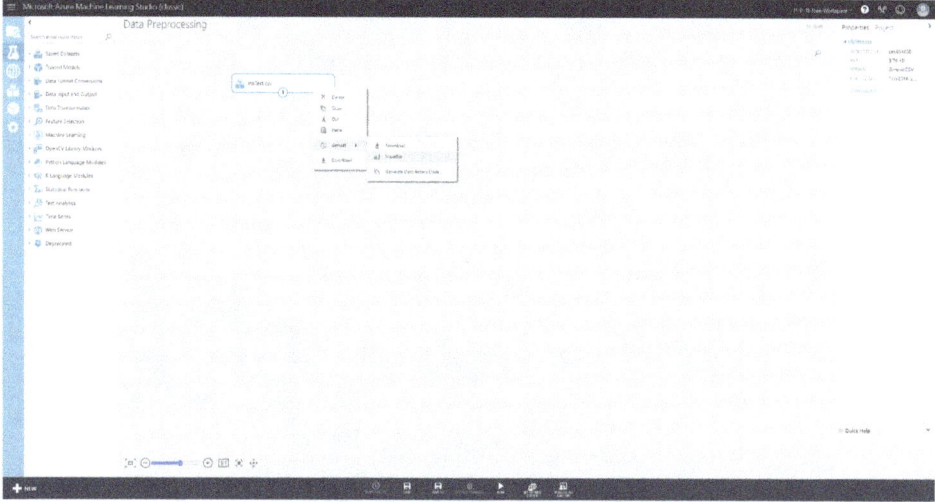

**Figure 6-31**    Method of dataset visualization

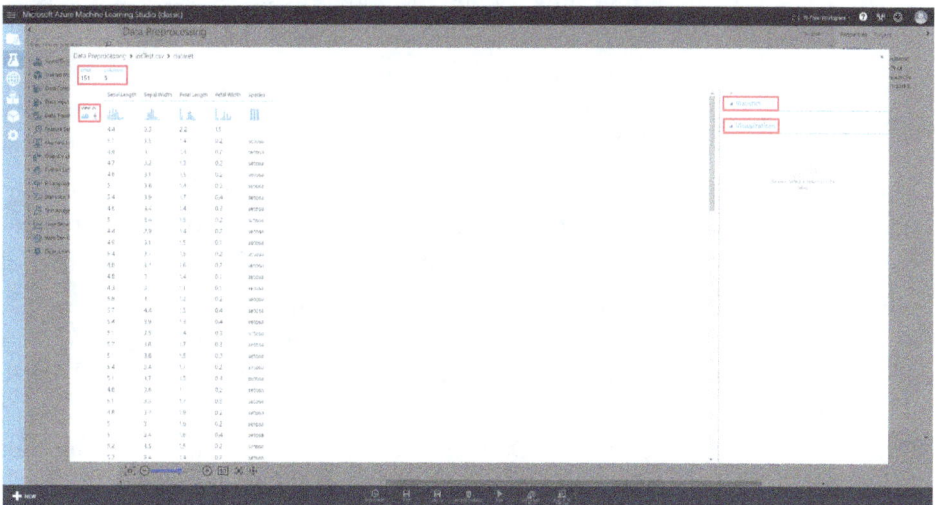

**Figure 6-32**    Result of dataset visualization

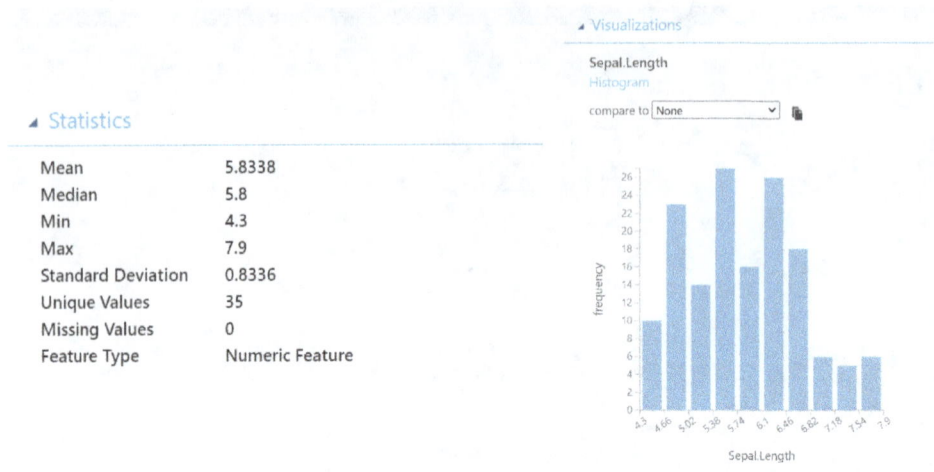

**Figure 6-33**    Statistics through data visualization

Also, you can check the statistics and histogram of the selected attribute as shown in Figure 6-33.

### 6.3.3 *Data Preprocessing*

Once you have inserted the experimental data and confirmed it through visualization, you should delete missing or unnecessary values. It is important to preprocess the data set before designing the experiment. Therefore, it is necessary to deal with missing values before applying the machine learning algorithm.

For the preprocessing practice, the row missing an attribute for "Species" was added, as shown in Figure 6-34. If the species value is missing, we should delete all the rows.

Import the "Select Columns in Dataset" block from the palette and include all the columns, as shown in Figure 6-35, using "Launch column selector".

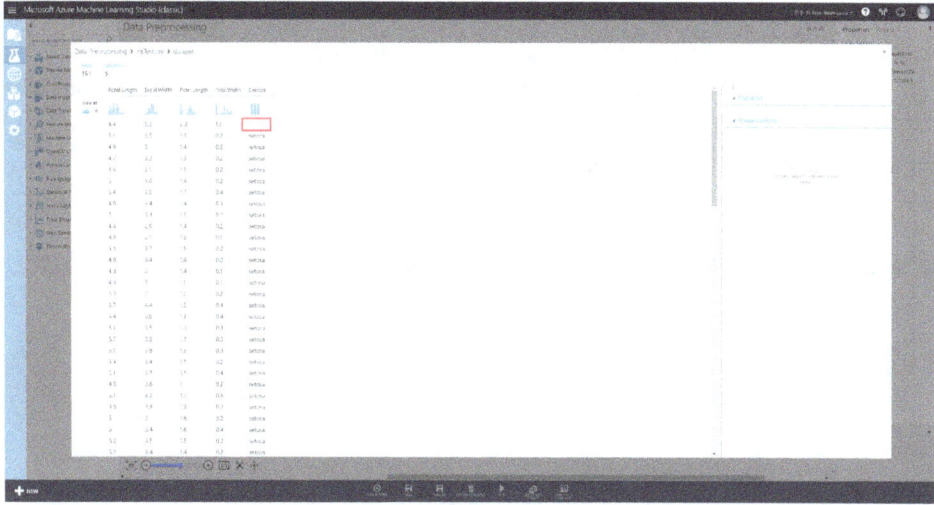

**Figure 6-34** Missing value

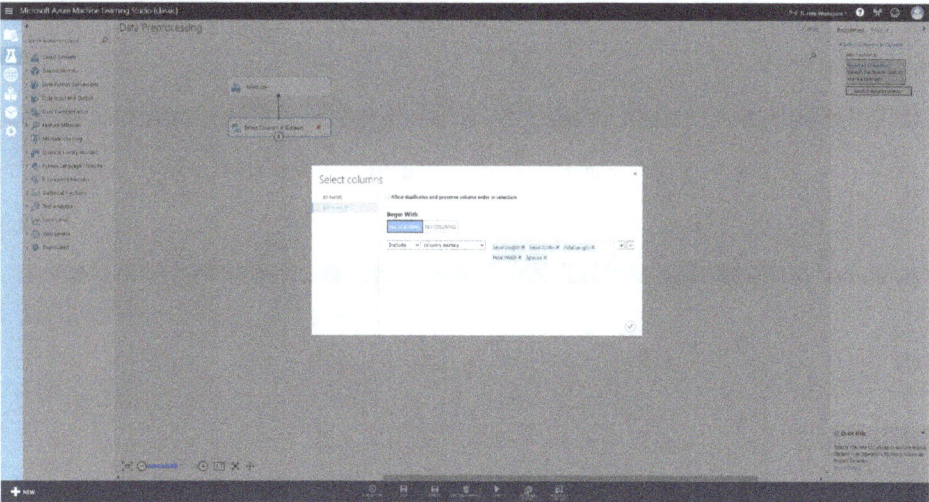

**Figure 6-35** Select Columns in Dataset to Inspect Missing Data

Import the "Clean Missing Data" block from the palette and set the cleaning mode to "Remove entire row", as shown in Figure 6-36.

After clicking the "Run" button and completing execution, right-click the "Clean Missing Data" block, as shown in Figure 6-37, and check the preprocessed data hovering over the "Cleaned Dataset" item. Then, click "Visualize".

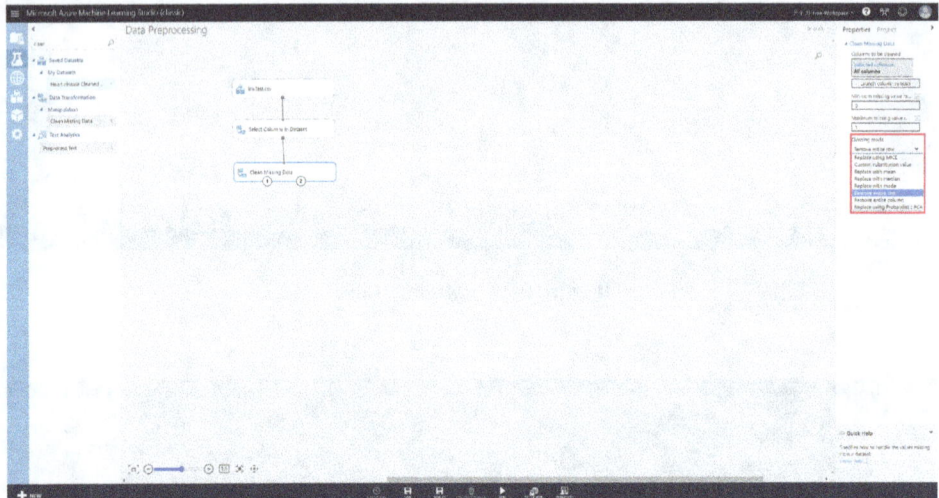

**Figure 6-36**   Missing data preprocessing using "Clean Missing Data"

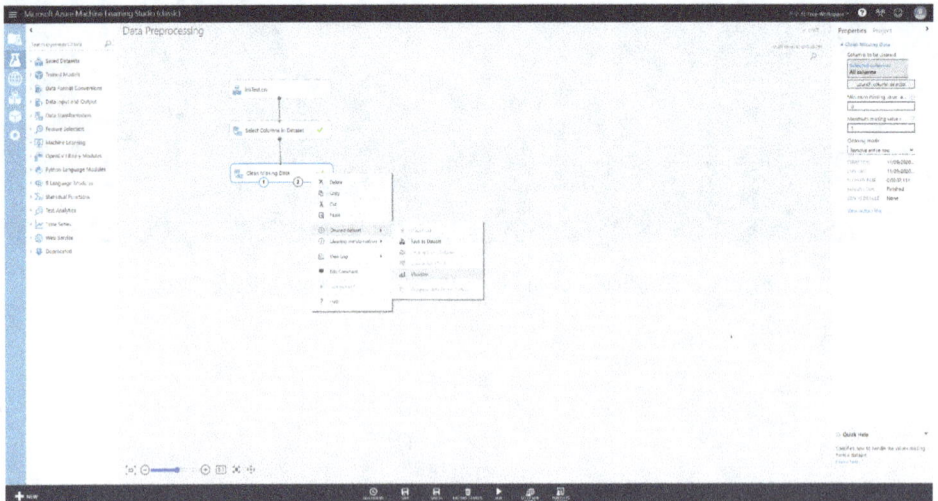

**Figure 6-37**   Preprocessed data visualization

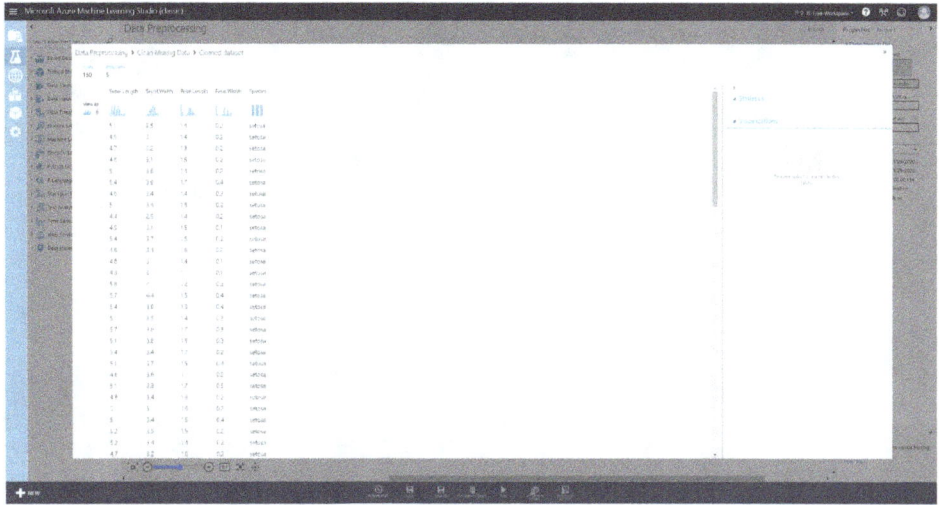

**Figure 6-38** Preprocessed data

As shown in Figure 6-38, if a value in the "Species" attribute is missing, confirm that the row has been deleted.

### 6.3.4 *Feature Definition*

Once you have dealt with missing values in the data, use the "Select Columns" in the "Dataset" block to select the attribute values in the data set in order to apply the algorithm. Of course, "Select Columns in Dataset" was used when preprocessing, but here it is used to select attribute values in order to implement the algorithm.

Place the "Select Columns in Dataset" block in the palette and include the Petal.Length, Petal.Width, and Species property values to apply the algorithm through the Launch Colum Selector, as shown in Figure 6-39. It is up to the data scientist to find the attribute values to implement the algorithm.

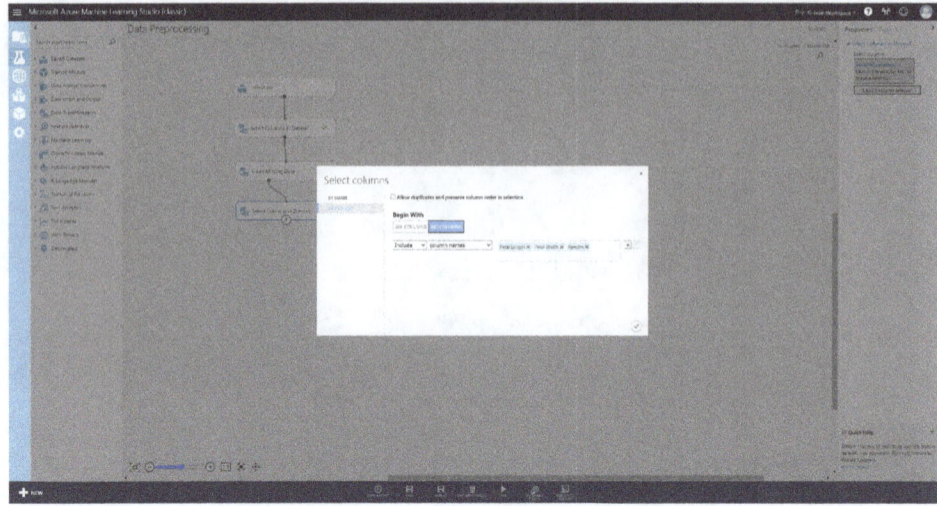

**Figure 6-39**   Select column for implementing algorithm

## 6.3.5 *Machine Learning Algorithm Selection and Implementation*

Once you have selected the attribute values to apply to the machine learning algorithm, you need to split the training and test data from the entire data set, and then select and implement the appropriate machine learning algorithm.

Place a "Split" block in the navigation palette, as shown in Figure 6-40, and use 70% as training data.

Let's import the Multiclass Decision Forest from the palette onto the canvas, place it, and then take a look at the Properties window. The algorithm to implement is the Multiclass Decision Forest. The Properties window makes it easy to modify the parameters of an algorithm. Proceed to the default value, as shown in Figure 6-41.

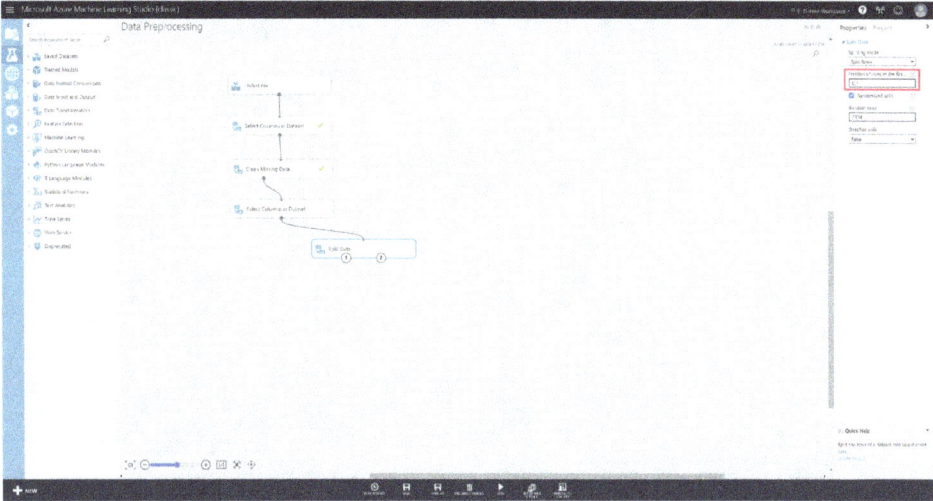

**Figure 6-40**   Data split

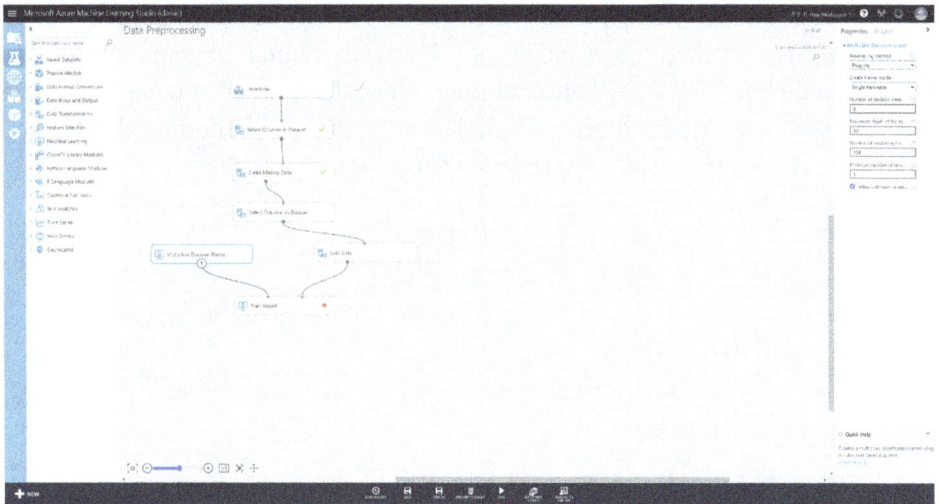

**Figure 6-41**   Property of Multiclass Decision Forest algorithm

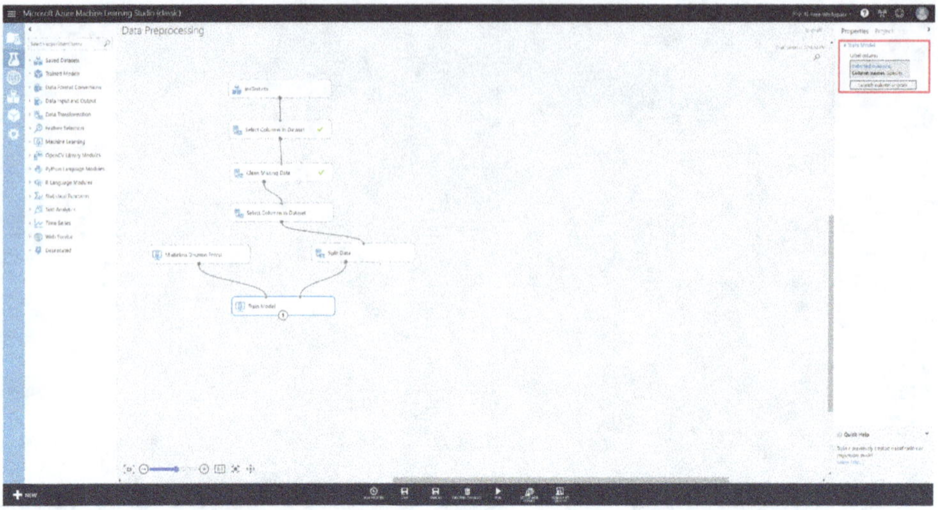

**Figure 6-42**   Label selection for training model

Place the Train Model block in the navigation palette as shown in Figure 6-42, select the Species that you want via the Launch Column Selector, connect as follows, and click the "Run" button to train. Once the training is complete, right-click on the Train Model block and click the "Visualize" button on the Trained model item to see the results. Figure 6-43 shows trained results.

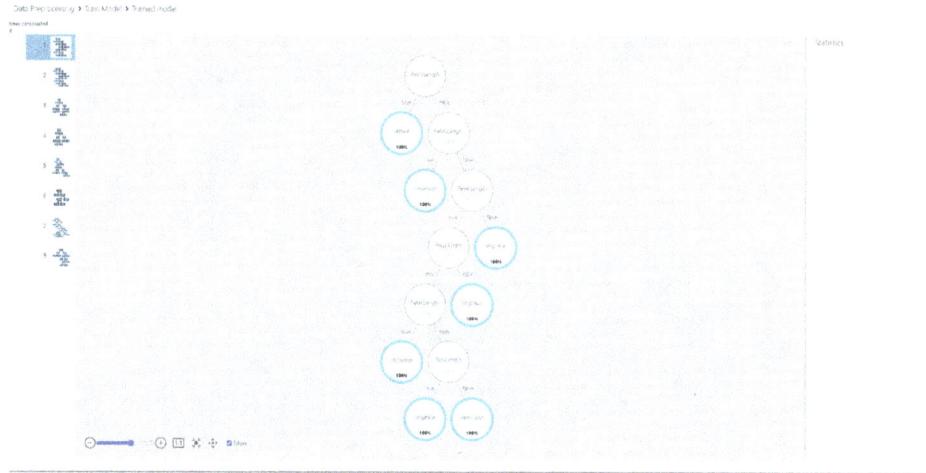

(a) Trained Tree 1

(b) Trained Tree 2

**Figure 6-43**   Trained results

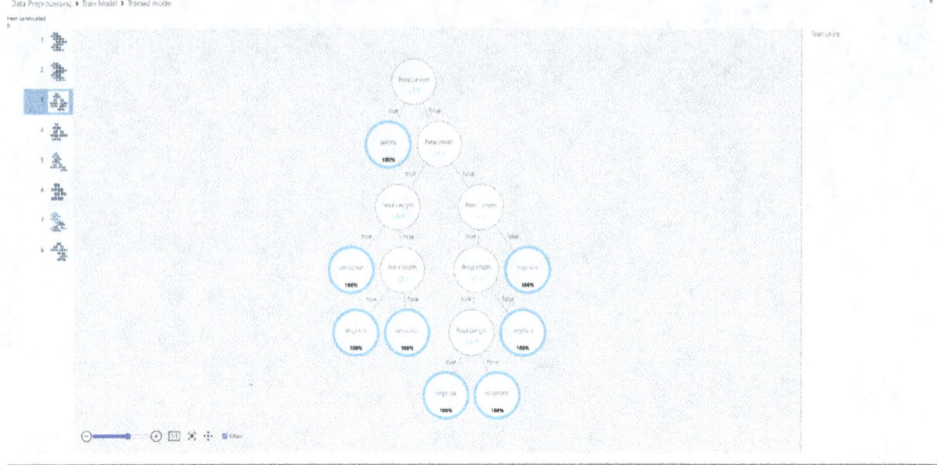

(c) Trained Tree 3

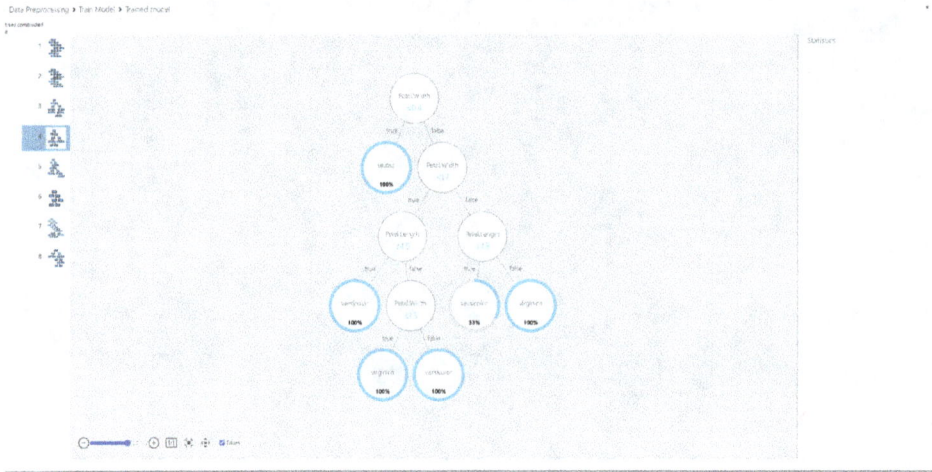

(d) Trained Tree 4

**Figure 6-43** (*Continued*)

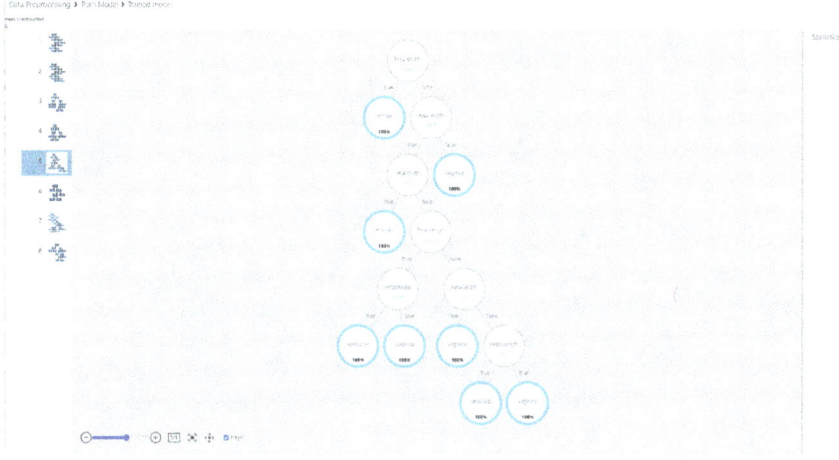

(e) Trained Tree 5

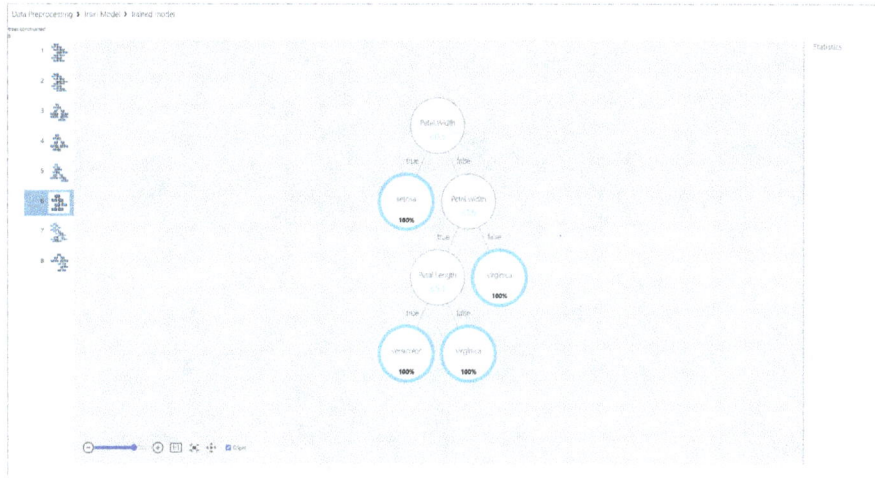

(f) Trained Tree 6

**Figure 6-43** (*Continued*)

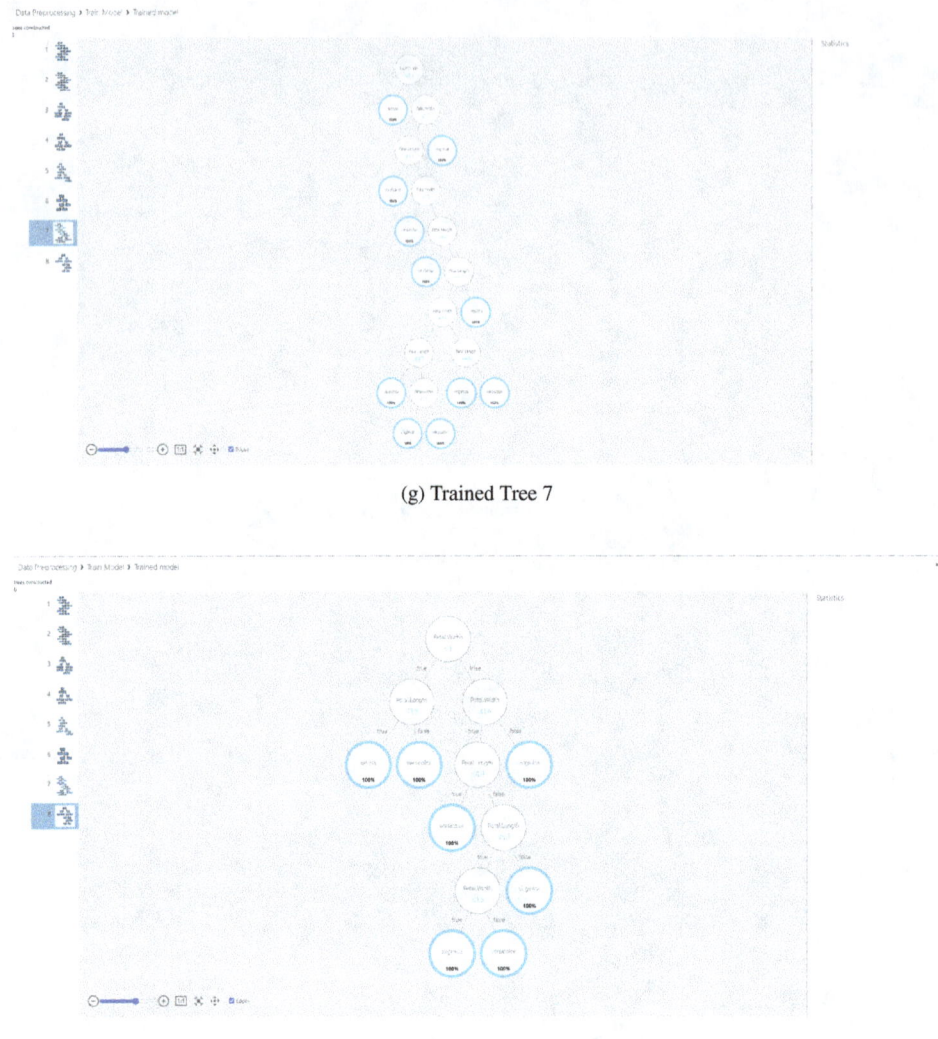

(g) Trained Tree 7

(h) Trained Tree 8

**Figure 6-43**   (*Continued*)

## 6.3.6 *Predicting with New Data*

Since the model has completed training, the remaining 30% of the test data can be applied to the trained model to evaluate how well it predicts.

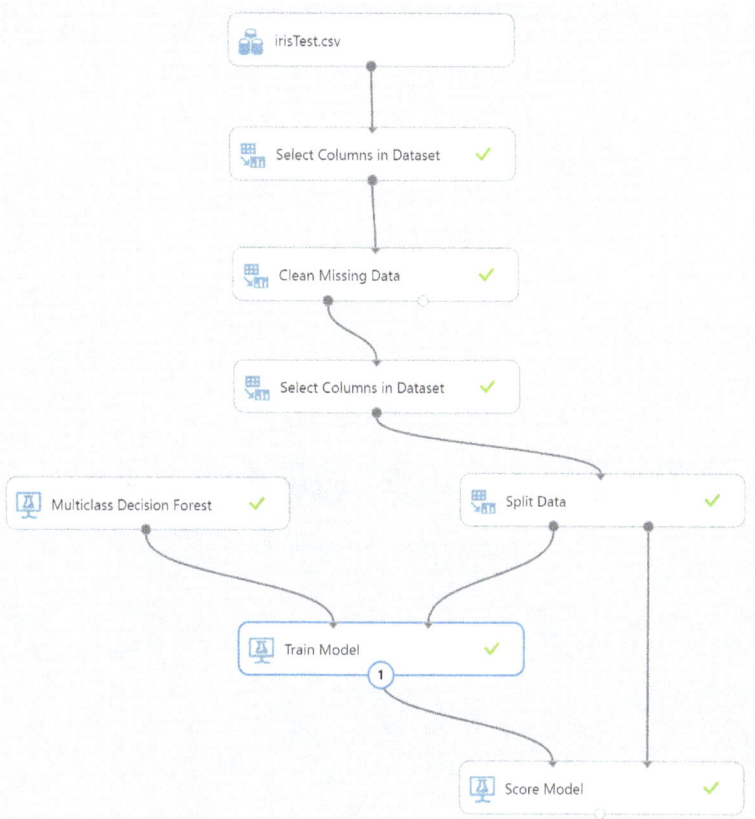

**Figure 6-44**  Arrangement for prediction model

Place the Score Model, as shown in Figure 6-44, connect the Train Model Block and the Split Block, and click the Run button to train. Let's check the types of irises predicted through visualization and the types of irises stored in the test data.

As shown in Figure 6-45, you can see that 30% of the test data has been applied to the prediction model.

To see the evaluation of the prediction results, add an "Evaluate Model" block to the canvas, and associate it with the "Train Model" block, as shown in Figure 6-46. Then press the "Run" button to run it. When it is done, check the results of the model evaluation with "Visualize".

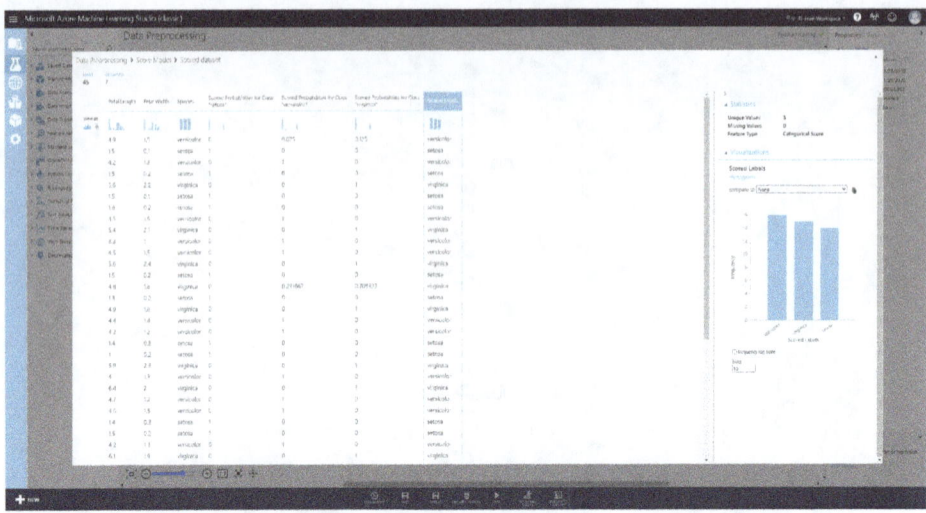

**Figure 6-45**   Predicted value

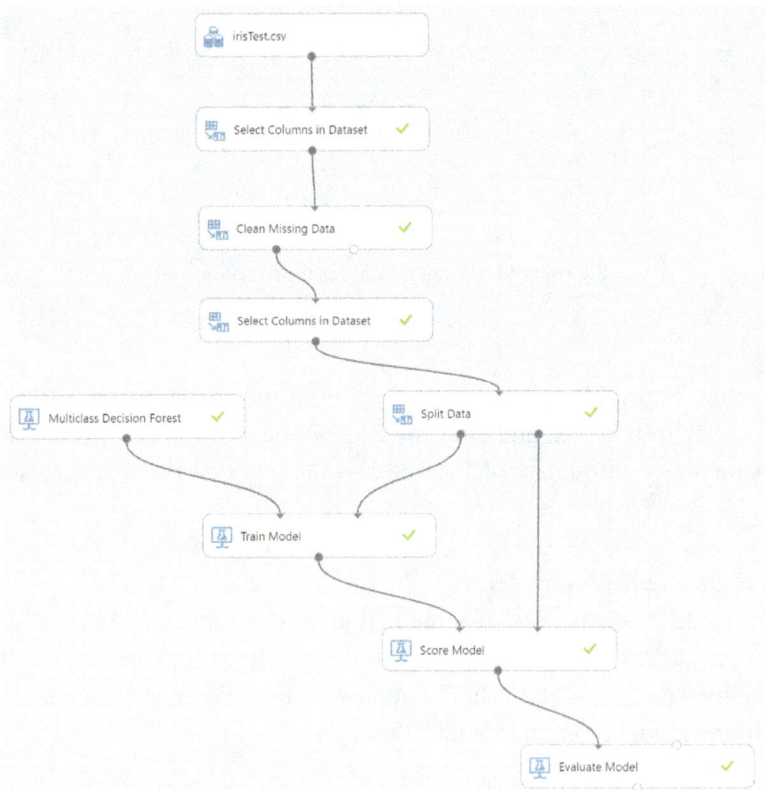

**Figure 6-46**   Arrangement for prediction model evaluation

As shown in Figure 6-47, the prediction model evaluation results show the accuracy, average precision, and average recall.

As shown in Figure 6-48, you can see the visualization by comparing the actual data with the predicted data.

◢ Metrics

| | |
|---|---|
| Overall accuracy | 0.977778 |
| Average accuracy | 0.985185 |
| Micro-averaged precision | 0.977778 |
| Macro-averaged precision | 0.979167 |
| Micro-averaged recall | 0.977778 |
| Macro-averaged recall | 0.977778 |

**Figure 6-47**    Prediction model evaluation results

◢ Confusion Matrix

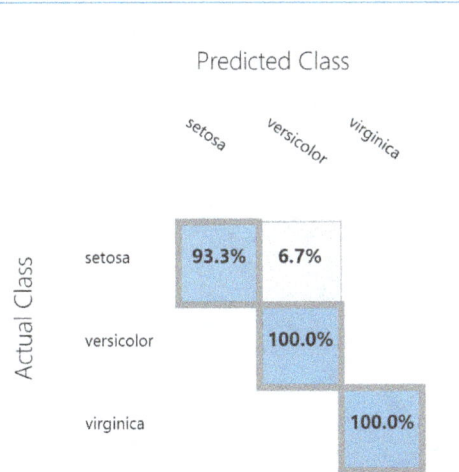

**Figure 6-48**    Visualization of prediction result

## 6.4 Practice Questions

Q1. Describe the process of creating a Prediction Model in Azure Machine Learning Studio


Q2. Describe the ROC (Receiver Operating Characteristic) Curve for model evaluation.


Q3. Describe the Accuracy, Precision, Recall, and F1 Scores for model evaluation.

# Chapter 7

# Create Prediction Models using Microsoft Azure Machine Learning Studio Web Service Deployment

## 7.1 Create Prediction Models using Microsoft Azure Machine Learning Studio Web Service Deployment Tutorial

Figure 7-1 shows the completion of the prediction model tutorial and suggests building a web service to use the prediction model remotely. Click the "Show Me" button to go through the tutorial to deploy the web service.

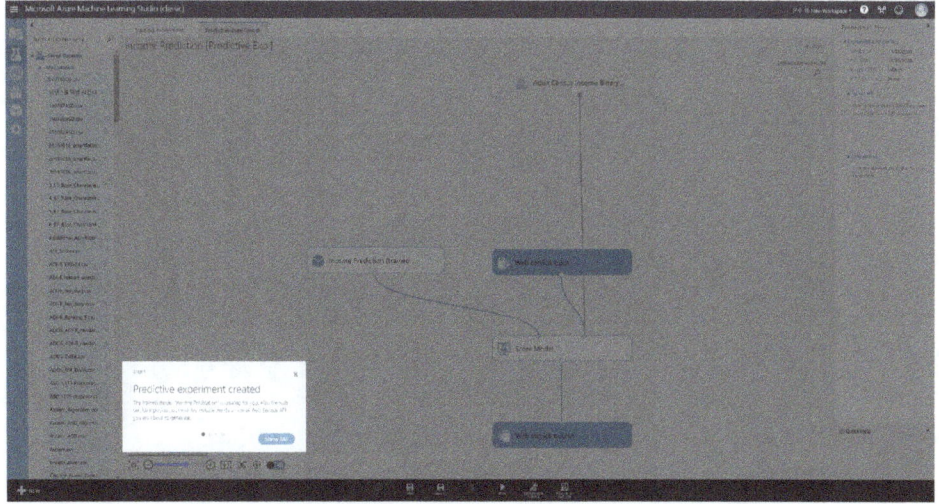

**Figure 7-1**  Web service deployment tutorial

Figure 7-2 shows "Predictive Web Service [Recommended]" being clicked to build the web service. Without any configuration, click to build automatically.

Figure 7-3 shows how the blocks are automatically placed, building a Web service input and output block.

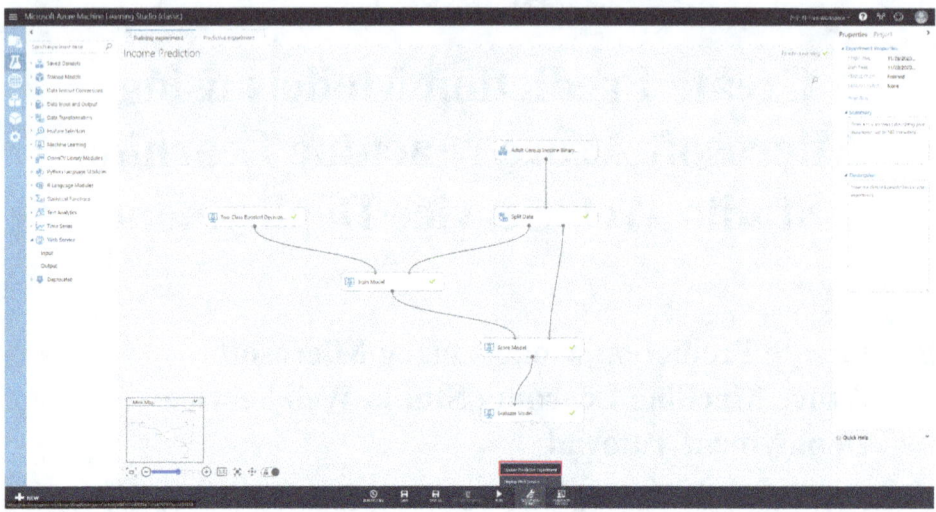

**Figure 7-2**  Build a web service

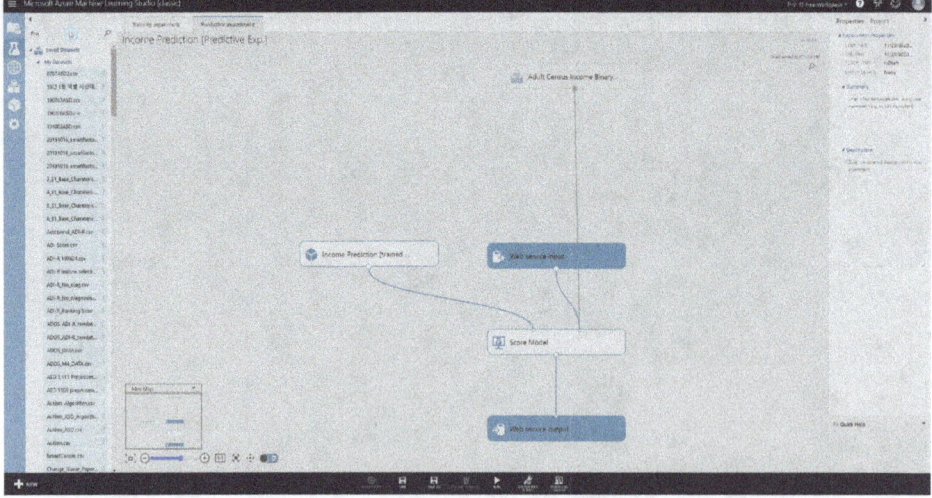

**Figure 7-3**  Canvas after building web service

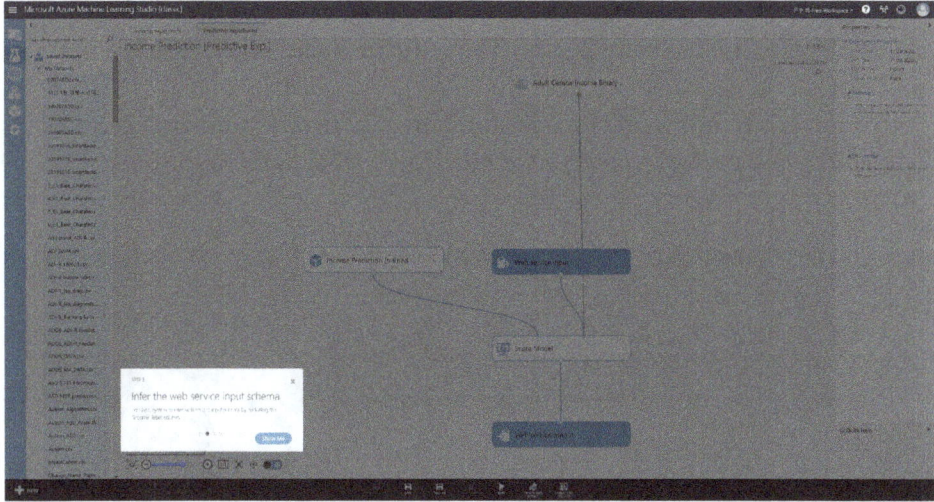

**Figure 7-4**   Input value setting

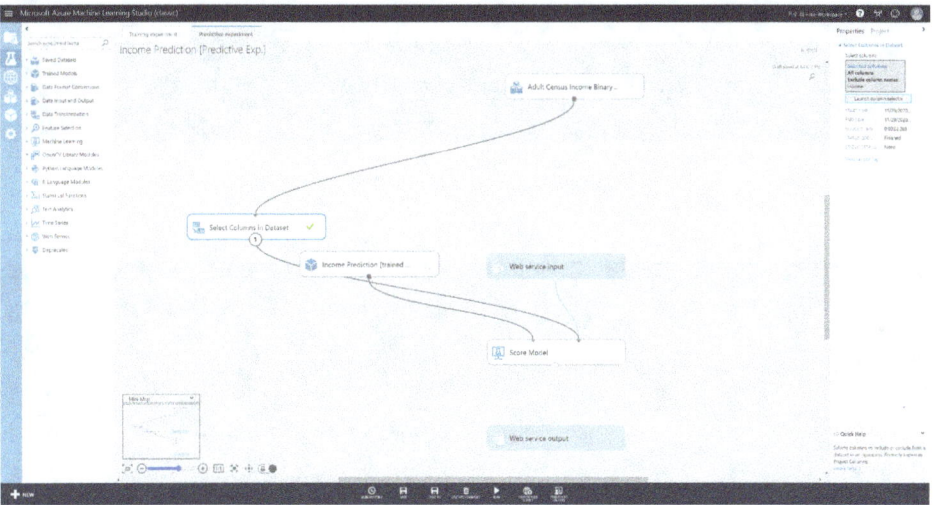

**Figure 7-5**   Place Select Column in Dataset for Prediction

Figure 7-4 shows the program suggesting that the user should define the Label Column input value as one other than the income attribute in order to predict the binary value income attribute value.

Place a "Select Columns in Dataset" block on the canvas and link it with the data set and Score Model, as shown in Figure 7-5. After that, click on "Launch Column Selector" to select a different label for prediction.

In Figure 7-6 the user must decide which attributes to input in order to make predictions through the web service. The attribute value is selected and checked by excluding the "Income" column from "ALL COLUMNS".

In Figure 7-7, the program suggests defining the output value as Scored Labels and Scored Probabilities to output the prediction value to the web service.

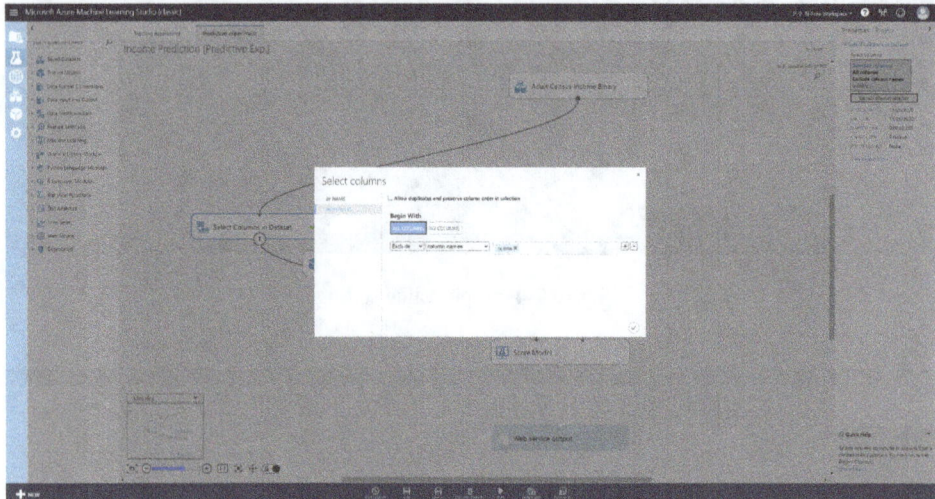

**Figure 7-6**    Select prediction value

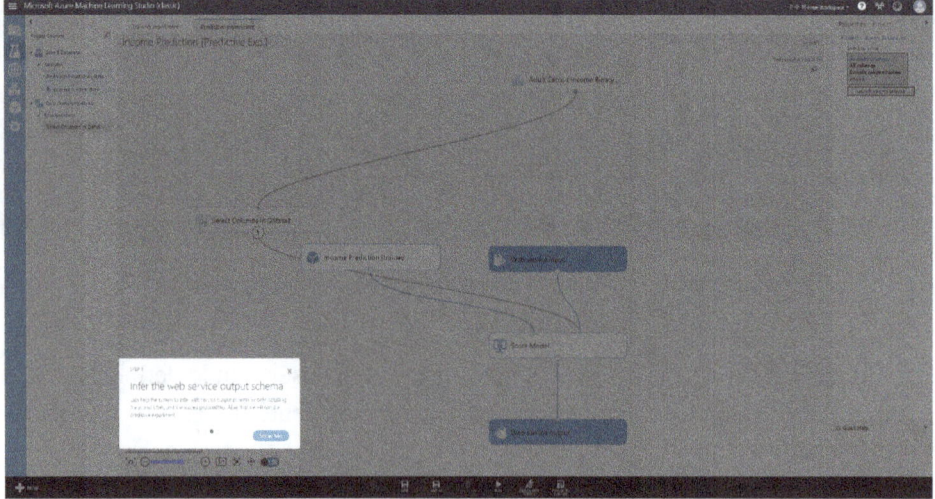

**Figure 7-7**    Prediction output setting

Place a "Select Columns in Dataset" block on the canvas, as shown in Figure 7-8, and connect the Score Model to the web service output. Then click on "Launch Column Selector" to select the Scored Labels and Scored Probabilities for output.

Select and include the "Scored Labels" and "Scored Probabilities" columns, beginning with "NO COLUMNS", as in Figure 7-9. Check it and click "Run" in the bottom menu to train your web service model.

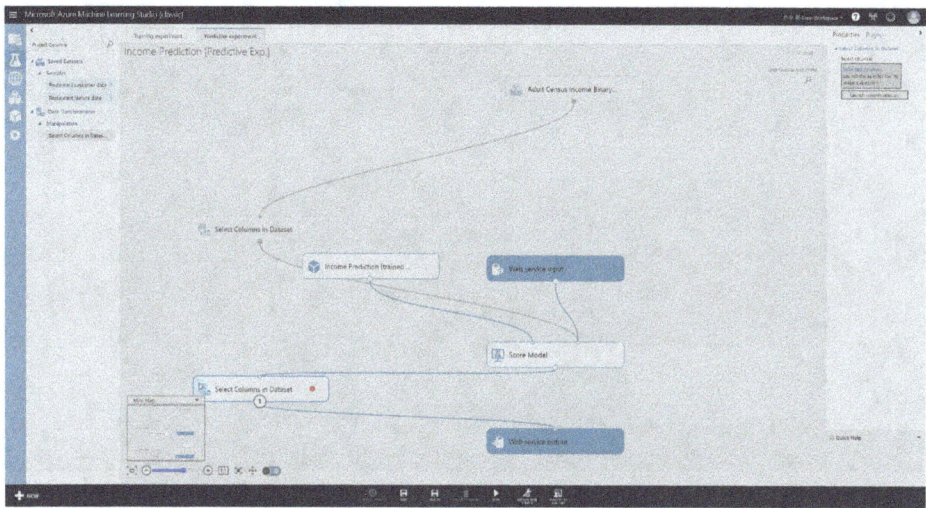

**Figure 7-8**   Arrangement for prediction value

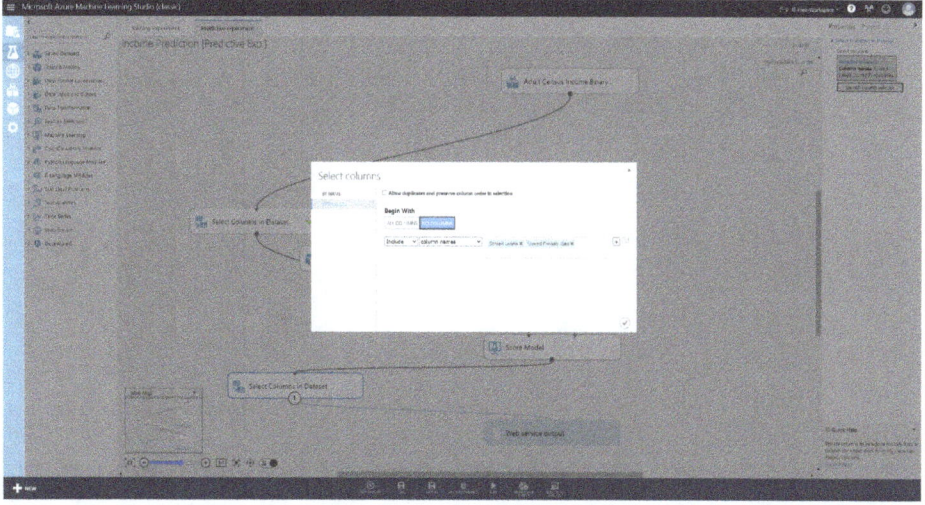

**Figure 7-9**   Prediction value and scored probabilities column selection

Once you have completed the training, as shown in Figure 7-10, you can use the completed web service. Click "Deploy Web Service" from the bottom menu.

Click the "Test" button in Figure 7-11 to predict the completed model.

Enter new data for prediction in the popup window, as shown in Figure 7-12, and click the check button.

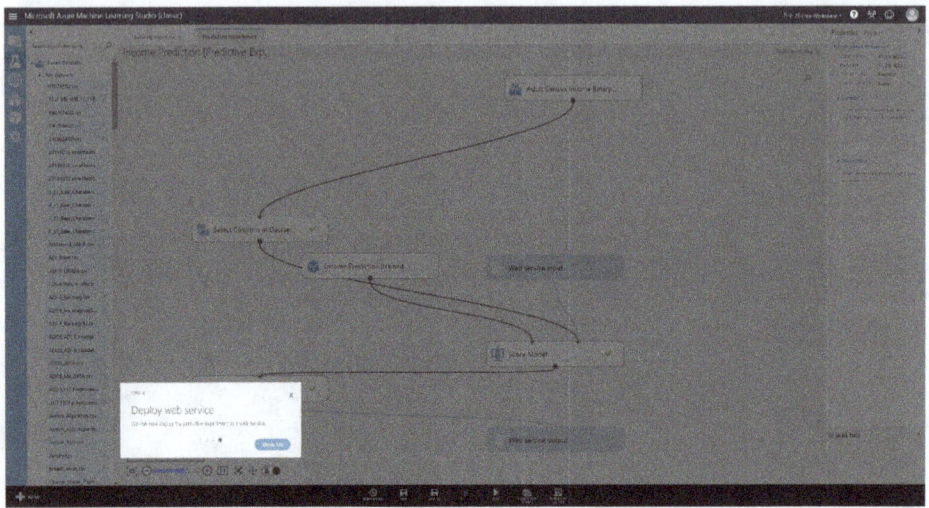

**Figure 7-10**   Web service deployment proposal

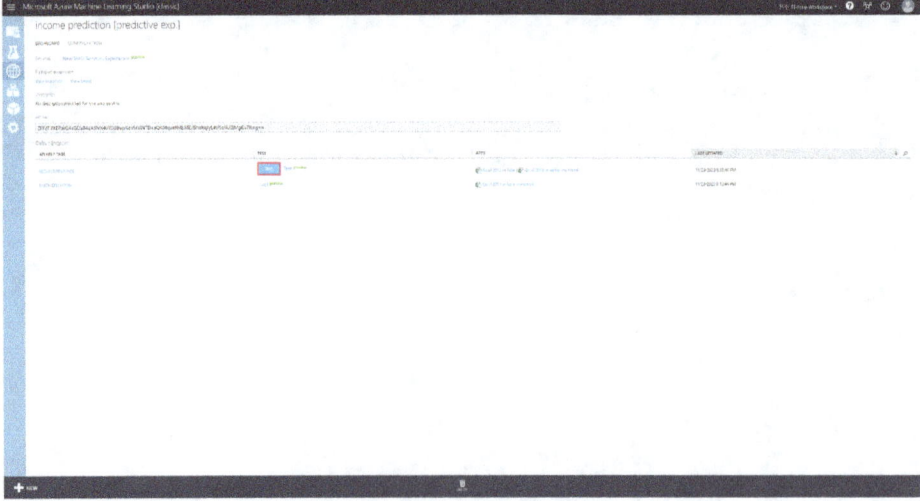

**Figure 7-11**   Use prediction model directly

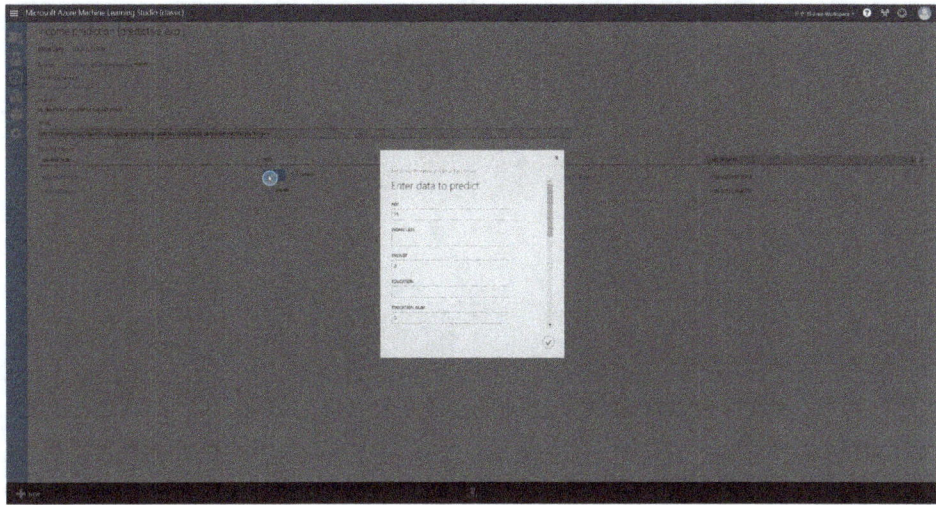

**Figure 7-12**   Enter data for predictive models when prompted by the popup window

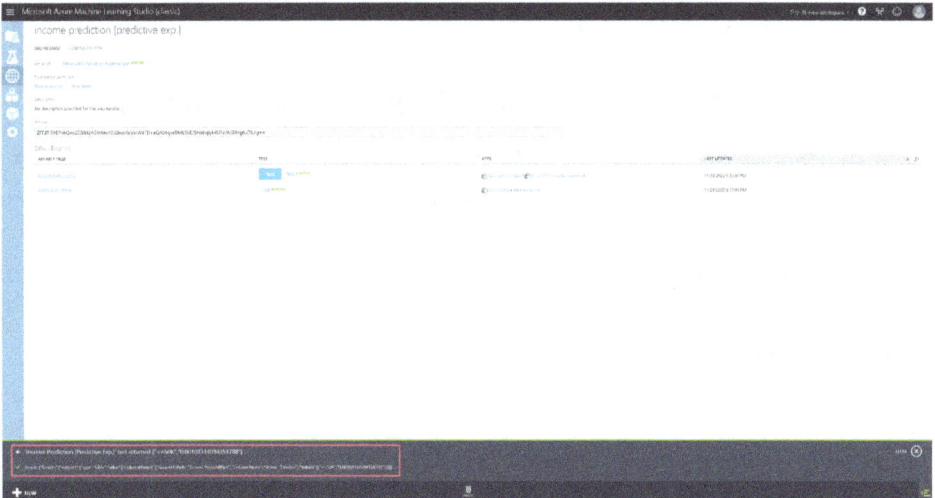

**Figure 7-13**   Prediction result

As shown in Figure 7-13, you can see the prediction result according to the input data at the bottom.

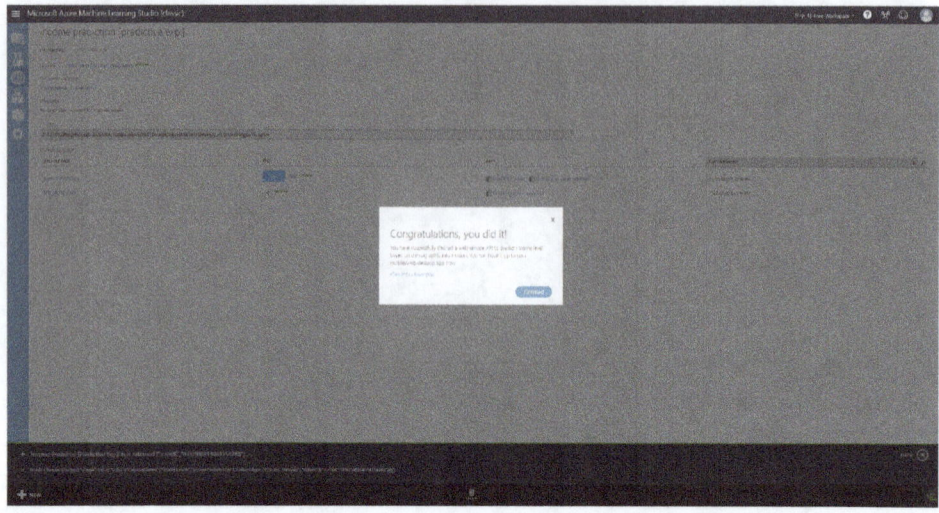

**Figure 7-14**　Web service deployment tutorial complete

Figure 7-14 shows the web service deployment tutorial is completed.

## 7.2　Web Service Model in the R and Python languages

Once the model is complete, it can be accessed from outside using the R and Python languages. The process of using a model as a web service will be explained through a tutorial.

If you have finished publishing the web service through the tutorial, you can see that it is in the web service category of "Income Prediction [Predictive exp.]", as shown in Figure 7-15.

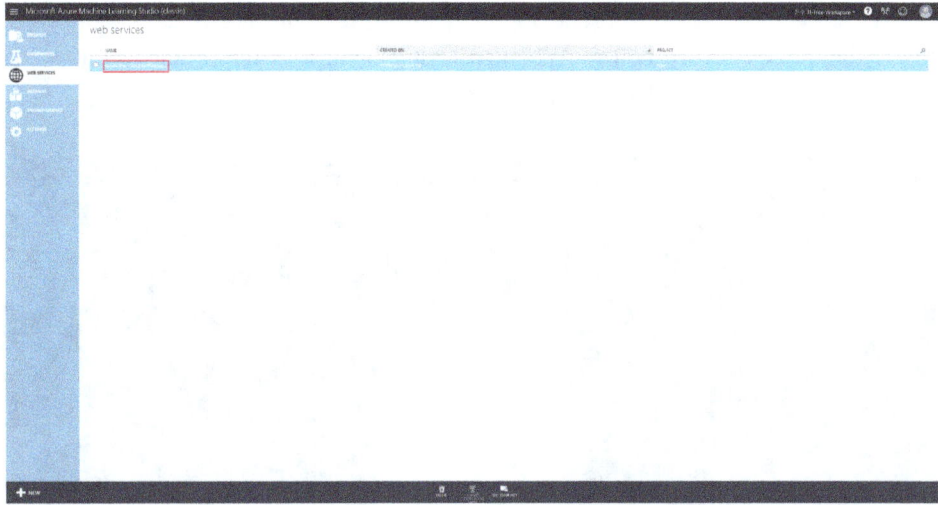

**Figure 7-15**　Web Service Created After Completing Tutorial

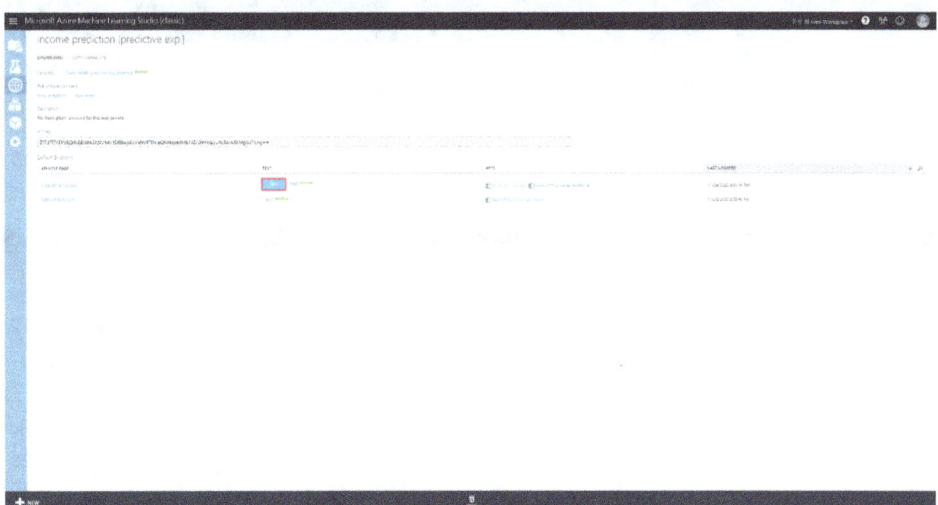

**Figure 7-16**　Web Service Dashboard

Before connecting R and Python, let's click the "Test" button in Figure 7-16.

As shown in Figure 7-17, a popup window for entering data is displayed. Insert the values shown below. When inserting data, it is important to know the type of data values used by the existing model.

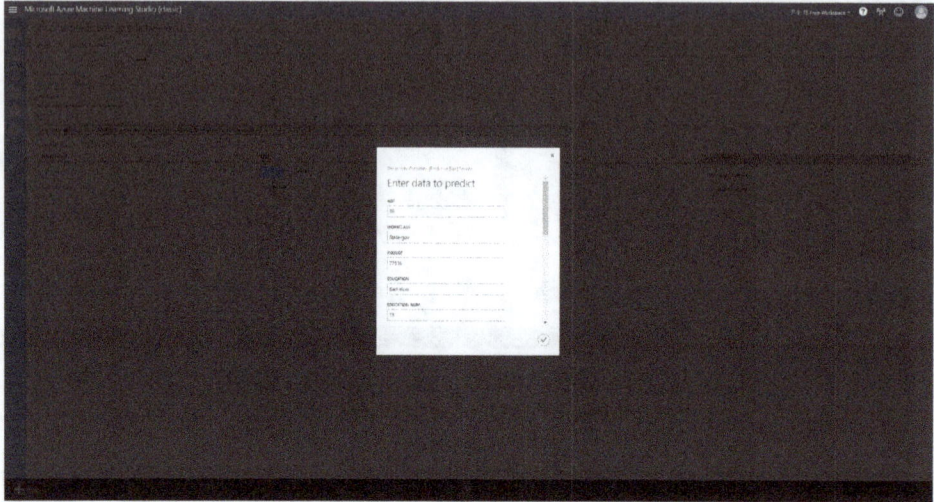

**Figure 7-17**   Data input popup window for prediction

Figure 7-18 shows the prediction without error.

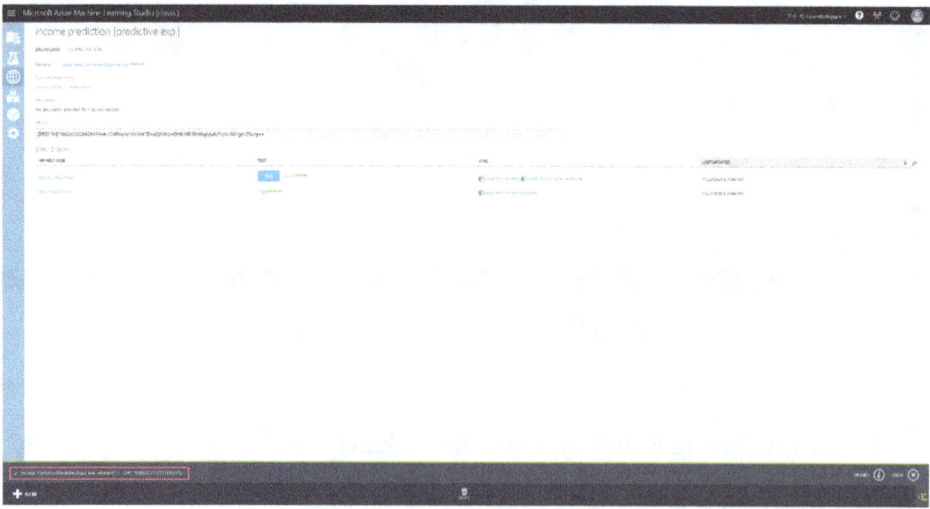

**Figure 7-18**   Prediction success

Click "New Web Service Experience" in Figure 7-19 to connect with R and Python.

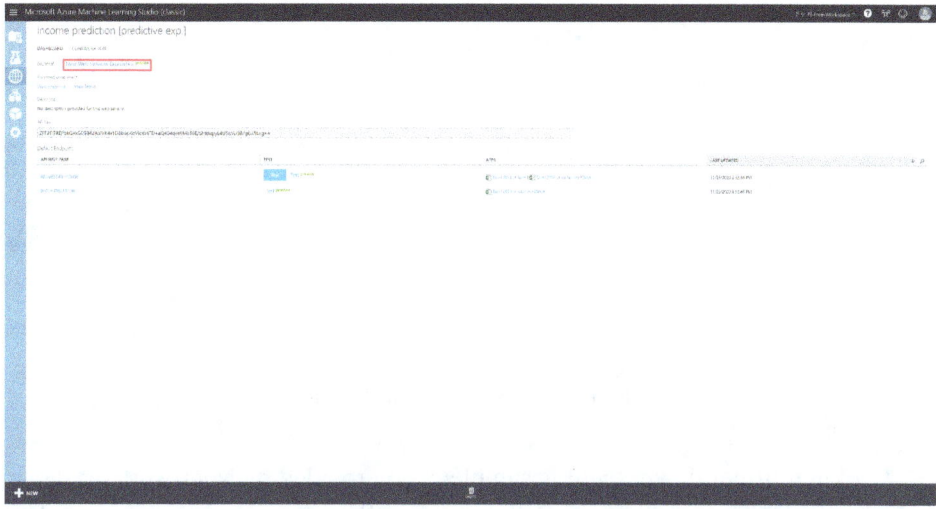

**Figure 7-19**   Web Service Dashboard

The loading window will pop up, and the categories will appear, as shown in Figure 7-20. Click "Use endpoint".

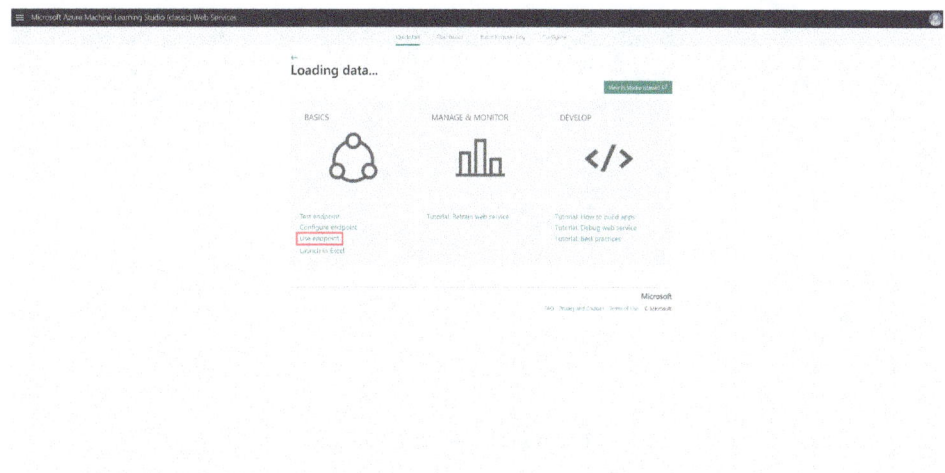

**Figure 7-20**   Web Service Management window

Figure 7-21 shows basic consumption information that provides various values of API Key. The keys provided here are essential, so it would be wise to save them.

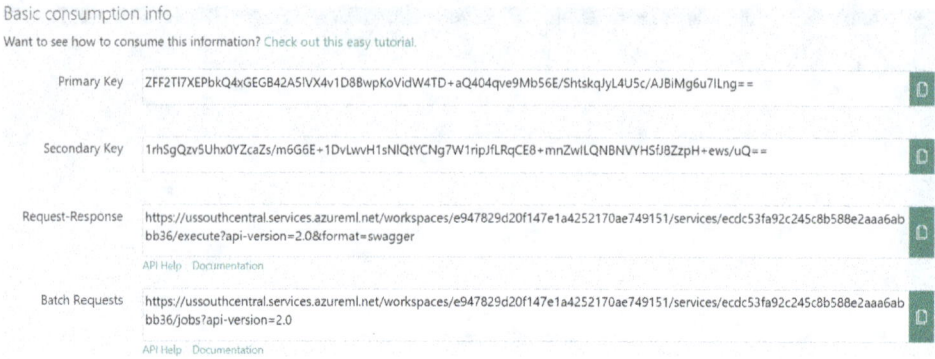

**Figure 7-21**　Basic consumption info

Looking further down, there is sample code provided by Microsoft, as shown in Figure 7-22. The C#, Python, Python3+, and R languages are supported. This book will work through Python 3+ and R.

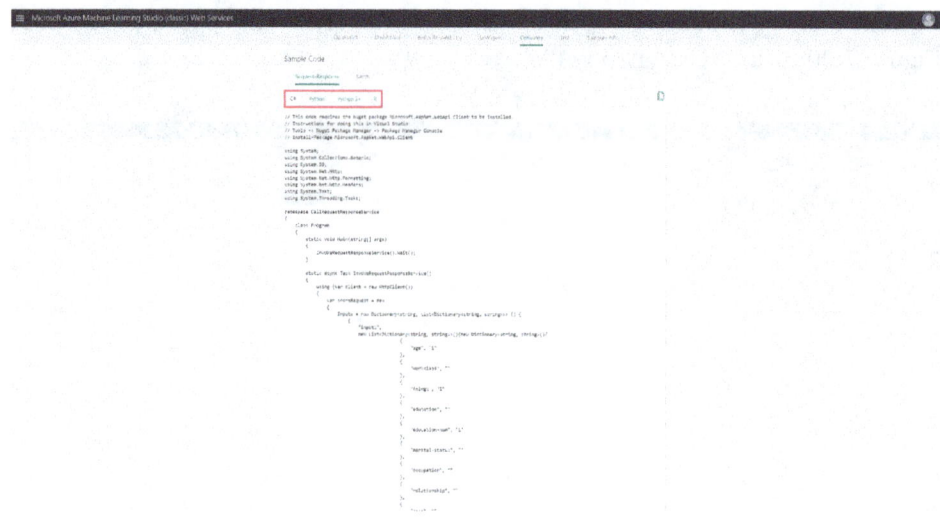

**Figure 7-22**　Sample code

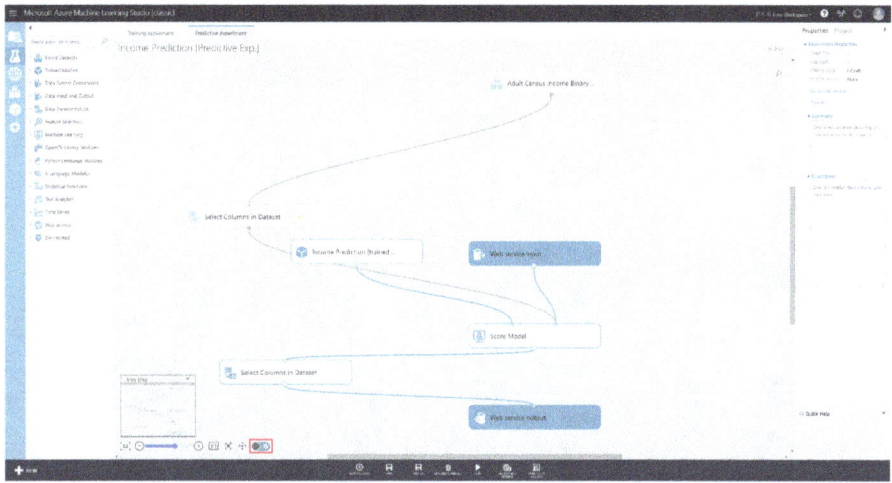

**Figure 7-23**   Click button to execute web services

Before connecting R and Python3 +, click the web service activation button, as shown in Figure 7-23.

## 7.2.1  *Integrating the Web Service with the R Language*

The R Studio program can be downloaded from the R Language official site, https://www.r-project.org/. The libraries used are bitops, RCurl, and rjson, which can be downloaded from R in Library Manager. To use the web service in R, Primary Key and Request-Respond values are required.

If you write and run the above code, you can see the prediction result from the model created in Azure Machine Learning Studio, as shown in Figure 7-24. You can use this to make your program more flexible.

```
> source("C:\\Users\\opxim\\Desktop\\azureWebService.R")
[1] "Result:"
$Results
$Results$output1
$Results$output1[[1]]
$Results$output1[[1]]$`Scored Labels`
[1] "<=50K"

$Results$output1[[1]]$`Scored Probabilities`
[1] "0.00355275371111929"
```

**Figure 7-24**   Result of connecting R

**Figure 7-25**   Azure Machine Learning Studio Web Services dashboard

When a value is returned from the external request, as shown in Figure 7-25, you can see the web service status in the dashboard.

### 7.2.2 *Integrating the Web Service with Python*

The official Python language (ver.3 and above) is available for download at https://www.python.org/. The libraries used are urlib.request and JSON. Python 3 can be used without downloading any libraries. To use a web service in Python, you need Primary Key and Request-Respond values, just as you would use a web service in R.

If you write and run the above code, you can see the predictions in the Python console due to the model created in Azure Machine Learning Studio, as shown in Figure 7-26. You can use Azure Machine Learning Studio to make your program more flexible.

**Figure 7-26**   Result of connecting Python language

## 7.3  Practice Questions

Q1. R and Python languages are used for the integration of the Web Service. Explain the advantages of integration.


Q2. R and Python languages are used for the integration of the Web Service. Explain the process of integration.

# Chapter 8

# Creating a Prediction Model using Microsoft Azure Machine Learning Studio Script Integration

## 8.1 R Script Integration

If you are using Azure Machine Learning Studio and want to apply an algorithm other than the given algorithm block, you can create and use an R script block.

### 8.1.1 *Viewing Data using R Script*

In the palette, retrieve the "Adult Census Income Binary Classification" dataset from the palette and place it on the canvas, as shown in Figure 8-1. Also, place the "Execute R Script" block.

Link the data set to the "Execute R Script" block, as shown in Figure 8-2, and click the button on the upper right.

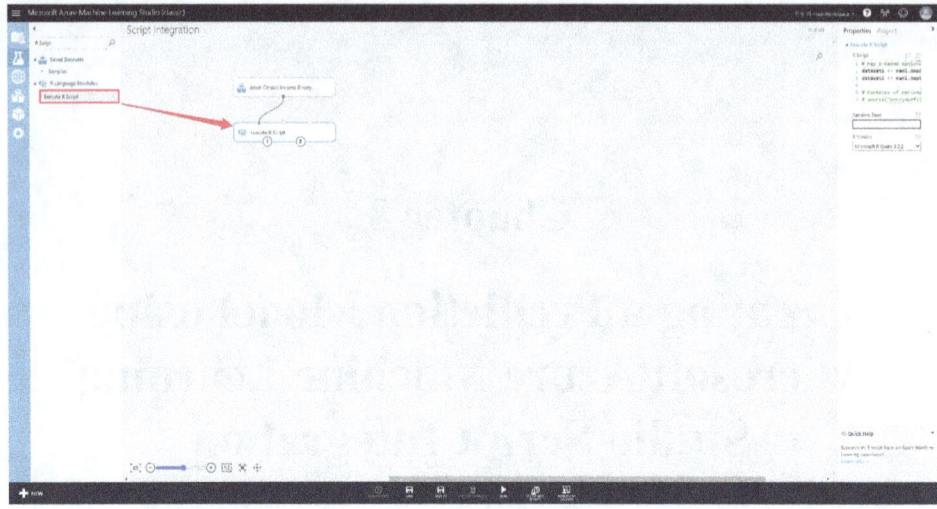

**Figure 8-1**   Arrangement for using R Script

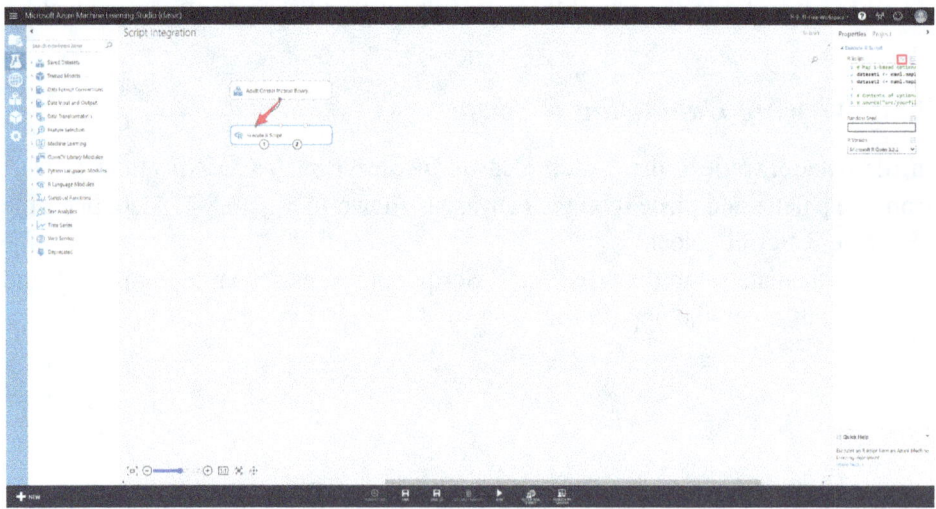

**Figure 8-2**   R Script modification

R Script in Azure Machine Learning Studio has two map Input ports. As shown in Figure 8-3, only one data set is connected so that data can be checked. Dataset1 contains the data connected to the first port.

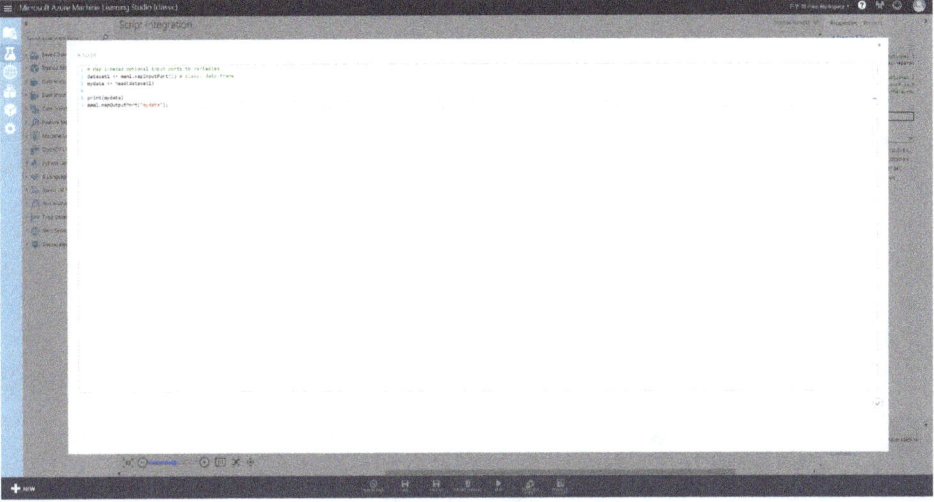

**Figure 8-3** R Script editing

## Source (8-3)

```
dataset1 <- maml.mapInputPort(1)
mydata <- dataset1[sample(1:nrow(dataset1), 50, replace=FALSE),]

data.set = mydata
print(mydata)

maml.mapOuputPort("data.set");
```

After that, take 50 dataset1 values and place them into "my data". You can check the entire sample data through the print function. The source can be found below.

Train the prediction model by clicking the "Run" button at the bottom. After that, you can check the extracted value by clicking "Visualize" in the "Result Dataset" menu, as shown in Figure 8-4.

Figure 8-5 shows that 50 dataset1 values are entered correctly.

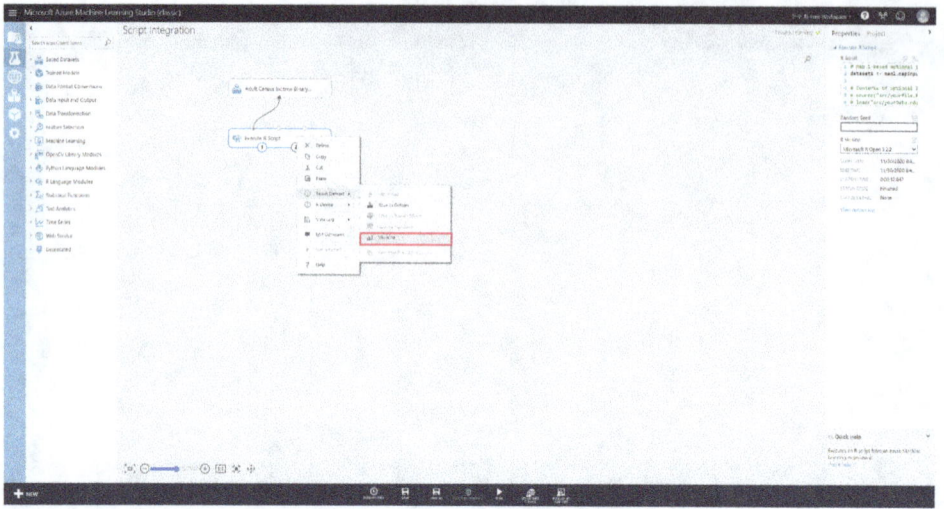

**Figure 8-4**　Data visualization using R Script

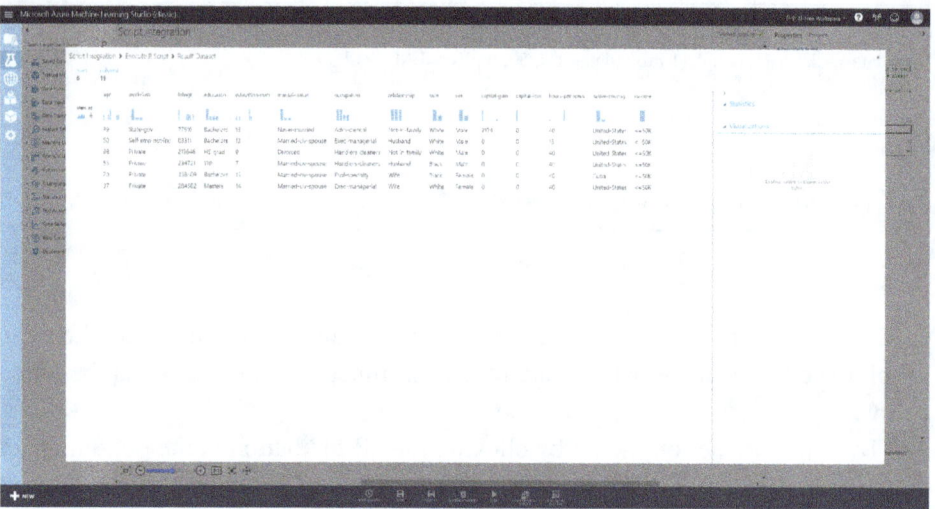

**Figure 8-5**　Result of data visualization

Figure 8-6 shows the part of data set extracted by the R script's head() function through the visualization of the R Device.

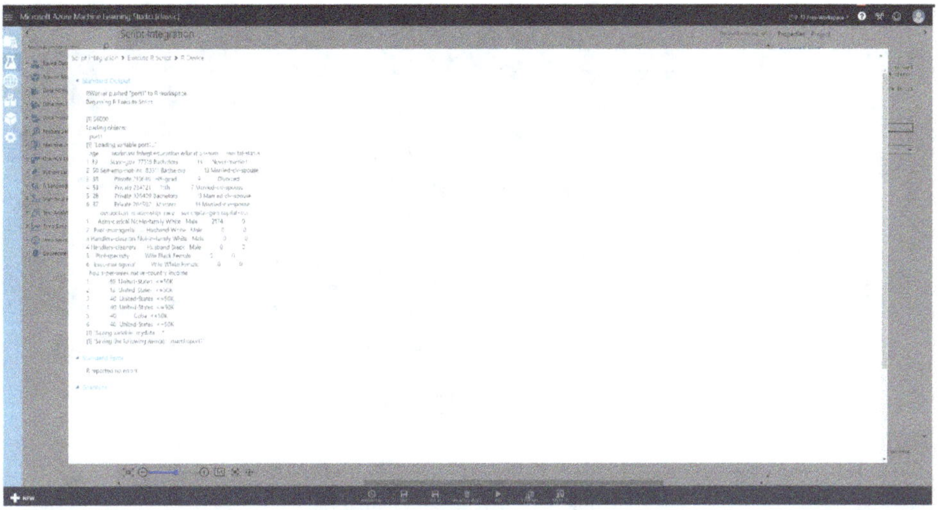

**Figure 8-6**  Data output result using R Script

## 8.1.2 *Implement Decision Tree using R Script*

Before applying the decision tree using R Script, let's look at the iris data set as example data. Iris data is provided by the statistician Fisher and summarizes the lengths of the sepal and petal for the three species of iris, Setosa, Versicolor, and Virginica, as shown in Figure 8-7. Iris data is easy to understand and is often used for machine learning due to the small amount of data in the data set.

Table 8-1 shows the description of the iris dataset. More information on the iris dataset can be found at the link https://archive.ics.uci.edu/ml/datasets/iris.

Before creating the model, import the iris data, as shown in Figure 8-8. Iris data is also provided as sample data in Azure Machine Learning Studio.

**Iris Versicolor**          **Iris Setosa**          **Iris Virginica**

**Figure 8-7**   Iris dataset

**Table 8-1**   Iris data explanation

| Column name | Mean | Data Type |
| --- | --- | --- |
| Species | One of three species of iris, Setosa, Versicolor, Virginica. | String |
| Sepal.Width | Iris's sepal width | Number |
| Sepal.Length | Iris's sepal length | Number |
| Petal.Width | Iris's petal width | Number |
| Petal.Length | Iris's petal length | Number |

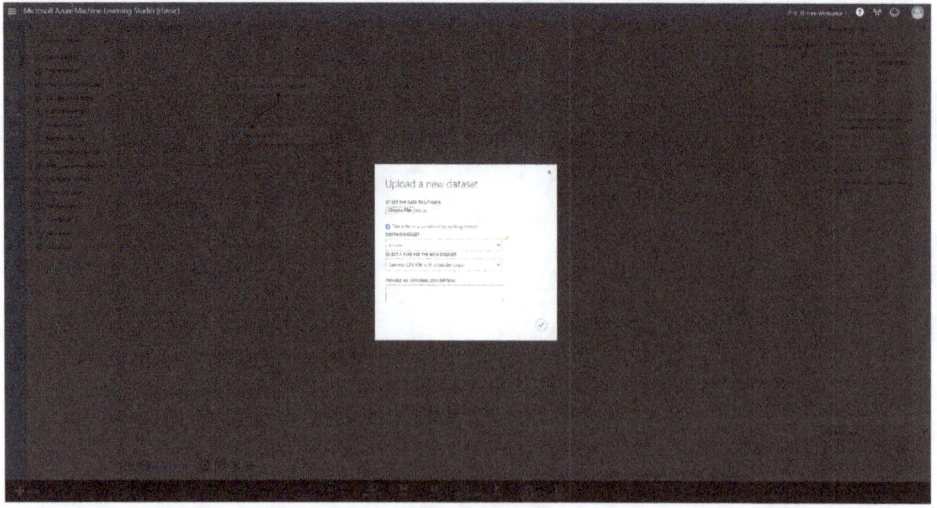

**Figure 8-8**   Importing data

After inserting the data set, place the data imported from the palette onto the canvas, as shown in Figure 8-9.

Place a "Select Columns in Dataset" block to define the columns in order apply the algorithm. After placing it as shown in Figure 8-10, click "Launch Column Selector".

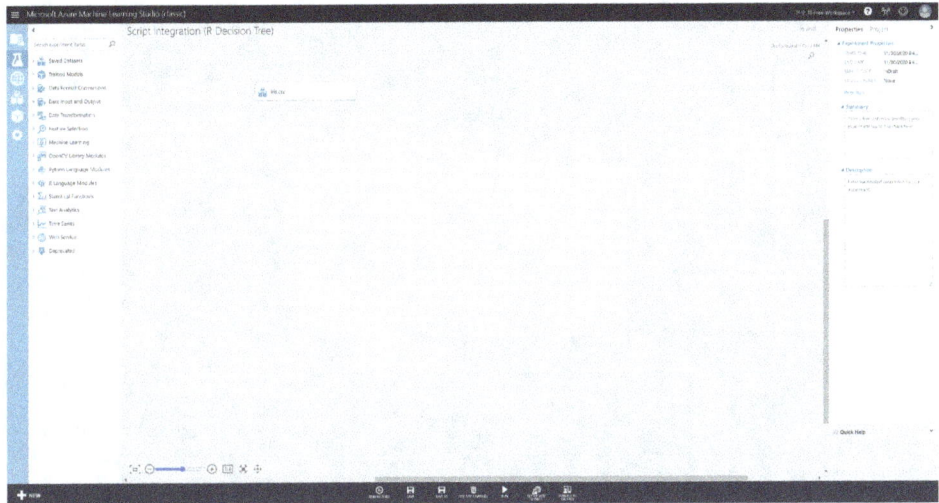

**Figure 8-9**  Place iris dataset

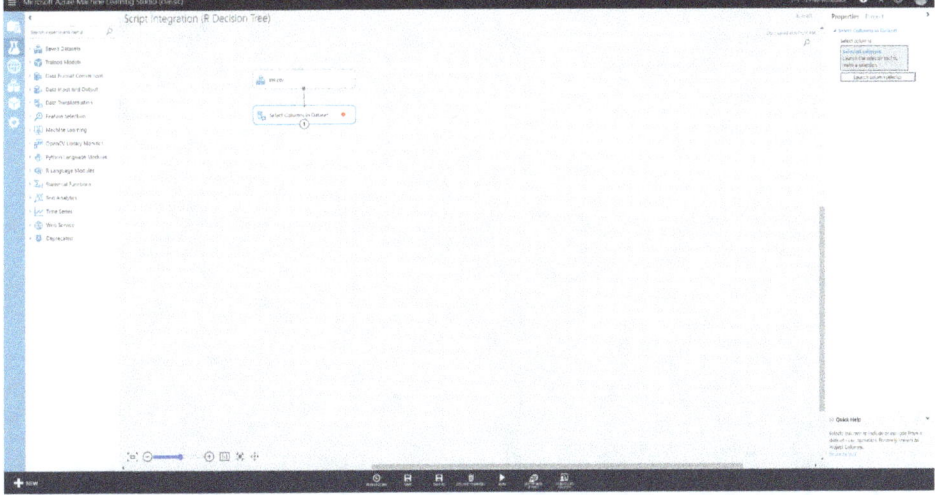

**Figure 8-10**  Arrange "Select Columns in Dataset" to implement the algorithm

Select all the columns to apply the algorithm, as shown in Figure 8-11.

Place the "Execute R Script" block, as shown in Figure 8-12, and click the small square in the R Script Properties window at the top right to modify the R script.

Enter the code at the bottom of the R Script window, as shown in Figure 8-13.

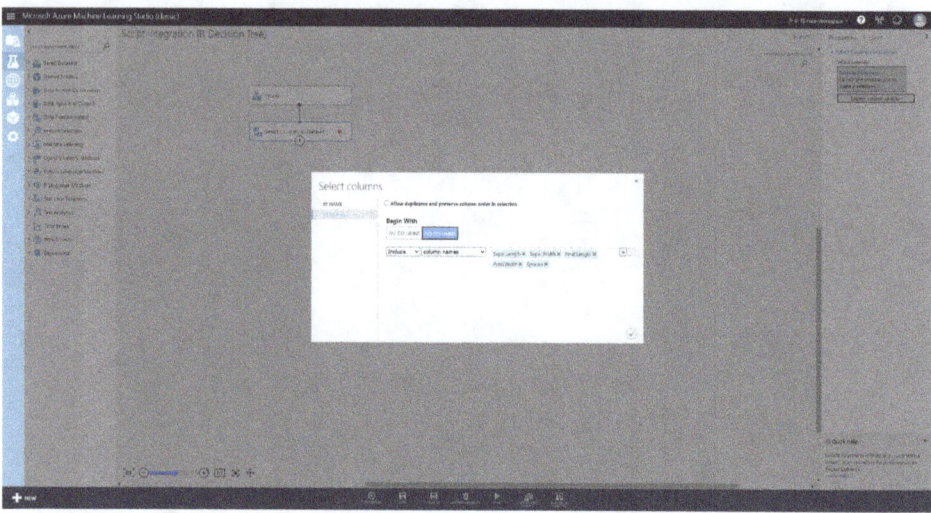

**Figure 8-11**    Select Columns for algorithm implementation

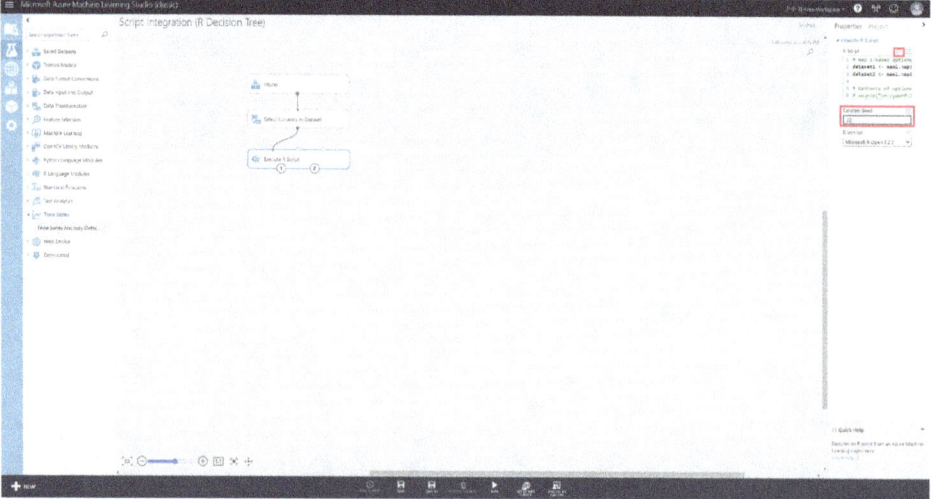

**Figure 8-12**    R Script modification

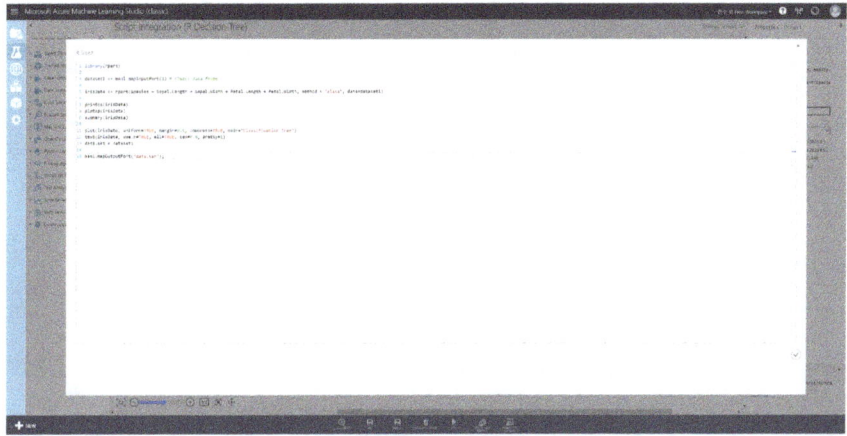

**Figure 8-13** R Script editing

## Source (8-13)

```
library(rpart)

Dataset1 <- maml.mapInputPort(1)
irisData <- rpart(Species~Sepal.Length + Sepal.Width + Petal.Length + Petal.Width, method="class",
data=Dataset1)

printcp(irisData)
plotcp(irisData)
summary(irisData)

plot(irisData, uniform=TRUE, margin=0.1, compress=TRUE, main="Classification Tree")
text(irisData, use.n=TRUE, all=TRUE, cex=0.8, pretty=1)
data.set = Dataset1

maml.mapOutputPort("data.set")
```

Use the visualization in Figure 8-14 to verify that the results applied to the R script are intact.

You can see that there are no problems with the result, as shown in Figure 8-15. Are the results from R Studio and Azure Machine Learning Studio the same?

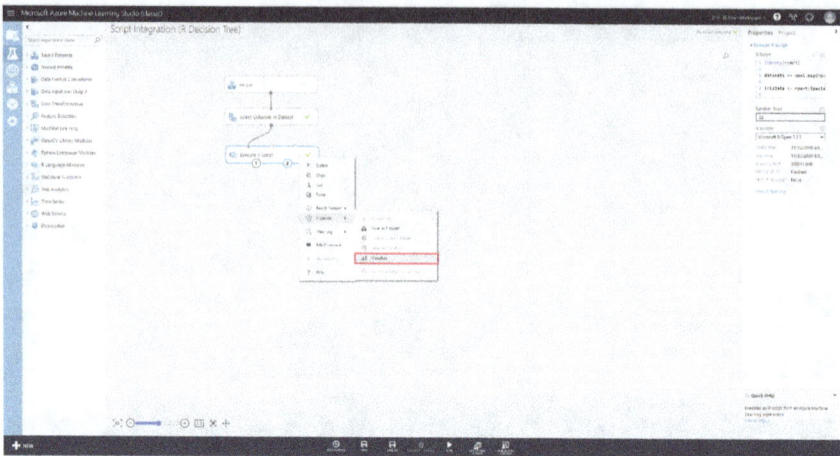

**Figure 8-14**   Visualization of decision tree result

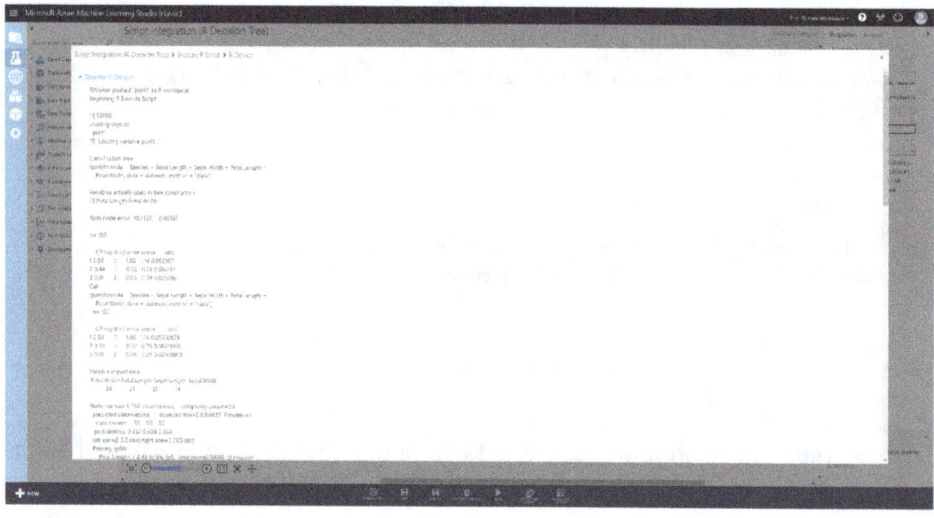

**Figure 8-15**   Decision tree result using R Script (Azure Machine Learning Studio)

As you can see in Figure 8-16, the same result has been applied to R Studio. Figure 8-17 shows the visualization of the decision tree applied to the iris data. Three species of irises can be seen classified according to the criteria.

```
Classification tree:
rpart(formula = Species ~ Sepal.Length + Sepal.Width + Petal.Length +
    Petal.Width, data = irisData, method = "class")

Variables actually used in tree construction:
[1] Petal.Length Petal.Width

Root node error: 100/150 = 0.66667

n= 150

    CP nsplit rel error xerror     xstd
1 0.50      0      1.00   1.24 0.046361
2 0.44      1      0.50   0.80 0.061101
3 0.01      2      0.06   0.09 0.029086
```

**Figure 8-16**   Decision tree result using R Script (R Studio)

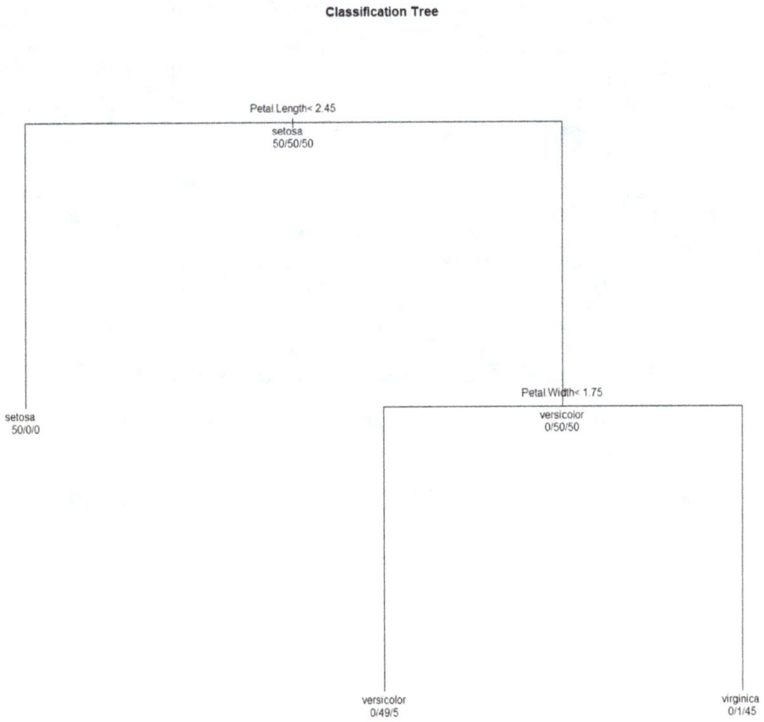

**Figure 8-17**   Decision tree visualization using R

## 8.2 Python Script Integration

Python, as well as R Script, can be integrated with Azure Machine Learning Studio.

Place and link iris data and click "Select Columns in Dataset" to integrate the R language. To integrate the Python language, select all columns to be included, as shown in Figure 8-18.

Then, import and place the "Execute Python Script" block from the palette, as shown in Figure 8-19.

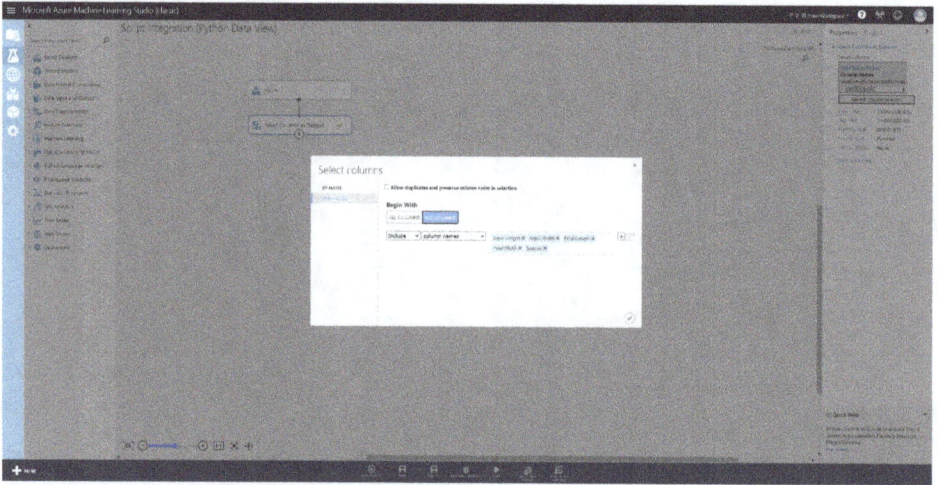

**Figure 8-18** Select Columns to implement algorithm

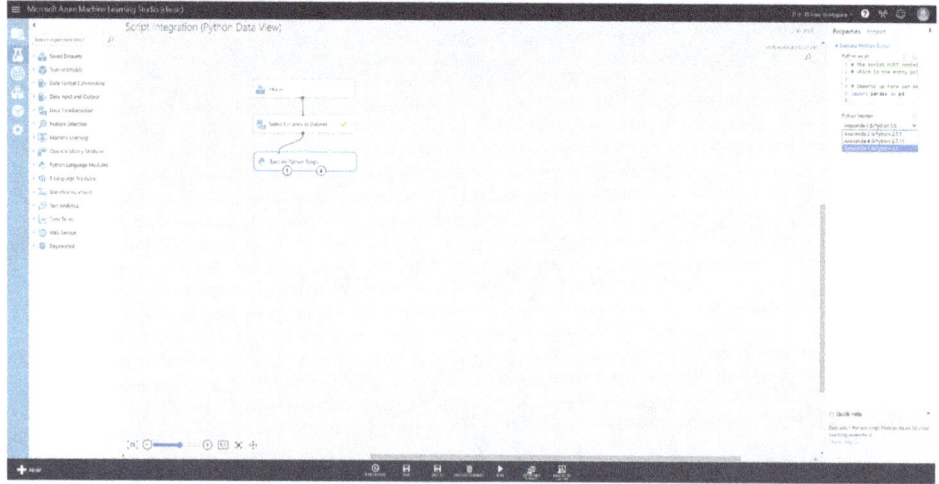

**Figure 8-19** Arrangement for Python integration

To modify the Python Script, click on the box in the upper right of the Python Script Properties window in Figure 8-20.

Refer to the usage of the "Execute Python" block in Figure 8-21 to understand how it works.

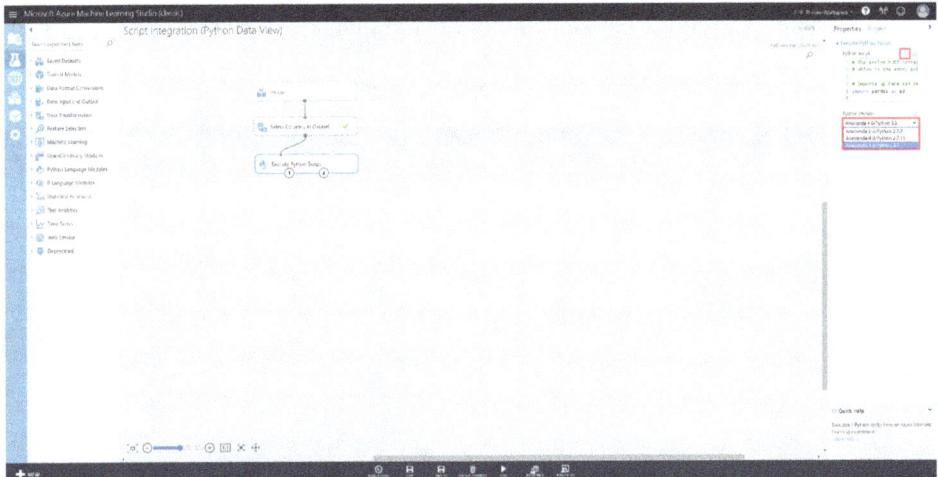

**Figure 8-20**   Python Script modification

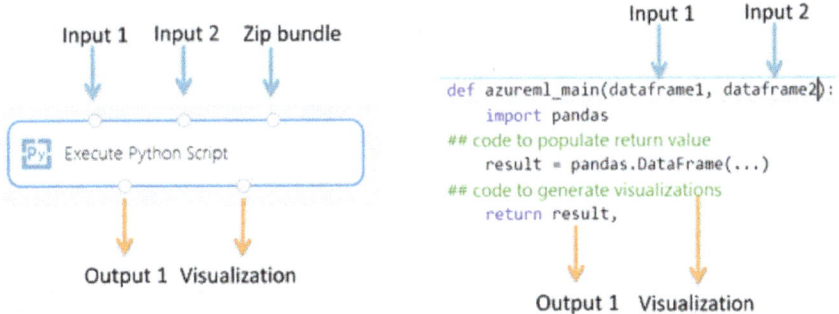

**Figure 8-21**   Execute Python block usage method

Enter the code at the bottom of the Python Script window, as shown in Figure 8-22.

When you have finished writing the Python script, click the "Run" button in the bottom menu to complete the learning. When done, right-click on the "Execute Python Script" block and then click "Visualize" on the Python Device item to see the result, as shown in Figure 8-23.

The result of Figure 8-24 shows that the data has been outputted correctly.

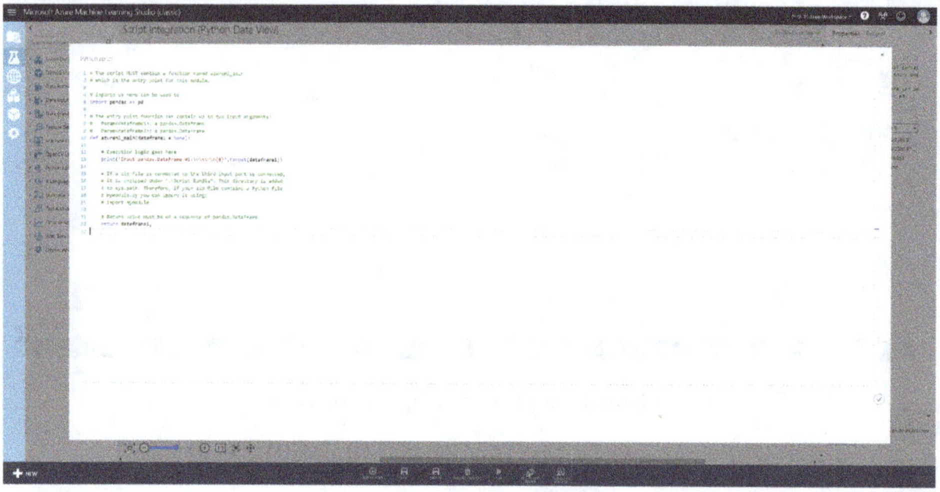

**Figure 8-22**  Python Script editing

## Source (8-22)

```
import pandas as pd

def azureml_main(dataframe1 = None):
        print('Input pandas.DataFrame #1: \r\n\r\n{0}'.format(dataframe1))
        return dataframe1,
```

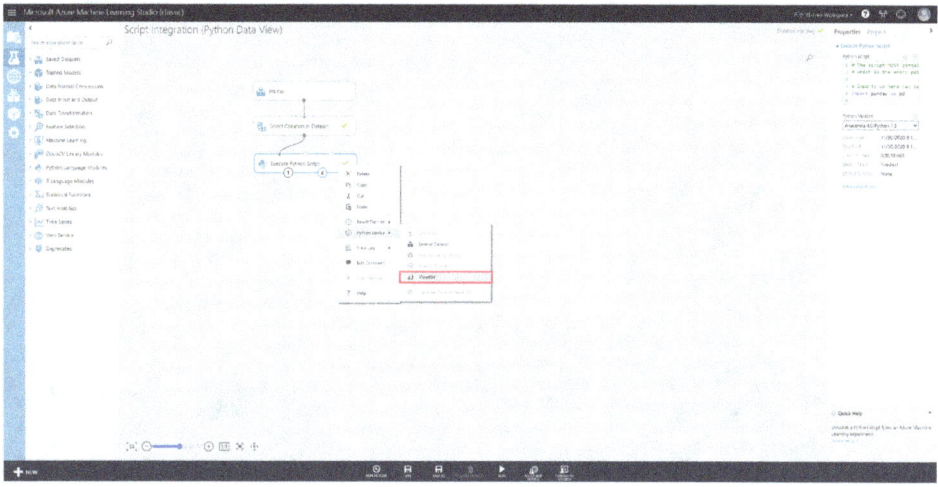

**Figure 8-23**    Data visualization using Python Script

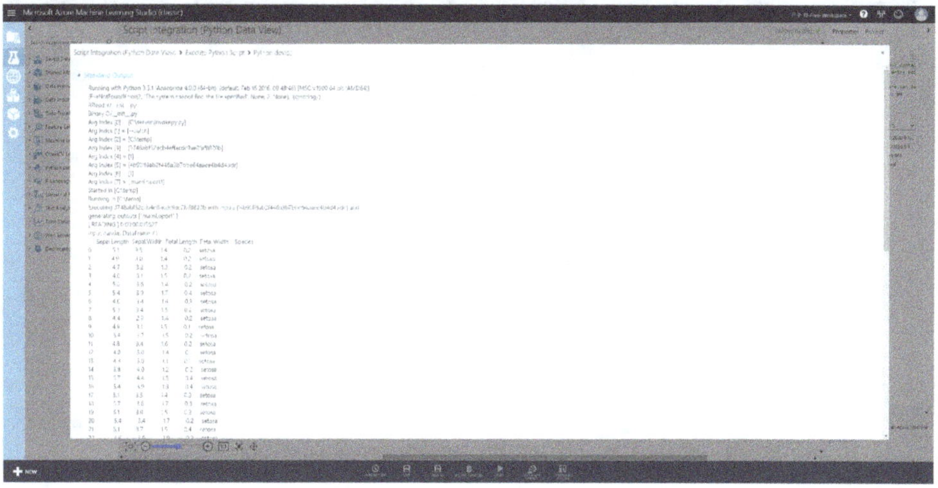

**Figure 8-24**    Data output result using Python Script

## 8.2.1 *Implement K-Means using Python Script*

Using the "Select Columns in Dataset" block, as shown in Figure 8-25, set the column to be sent to the "Execute Python Script" block.

Then import and place the "Execute Python Script" block from the palette, as shown in Figure 8-26.

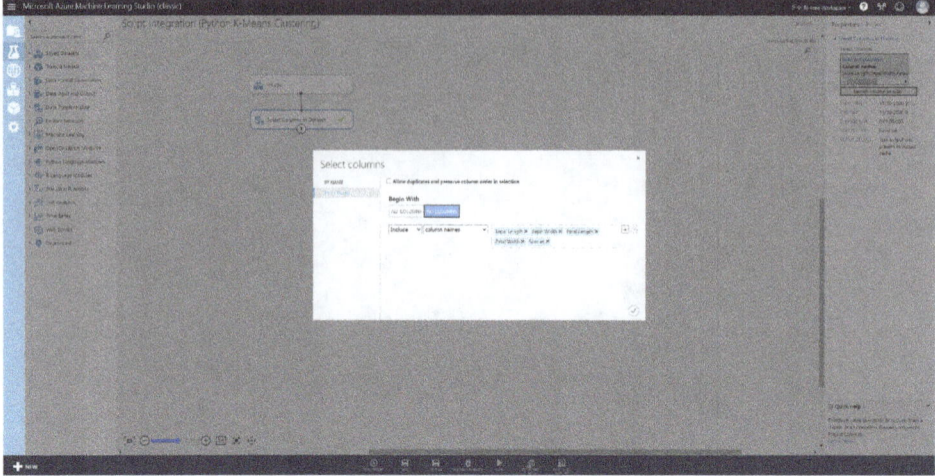

**Figure 8-25**    Select Columns to implement algorithm

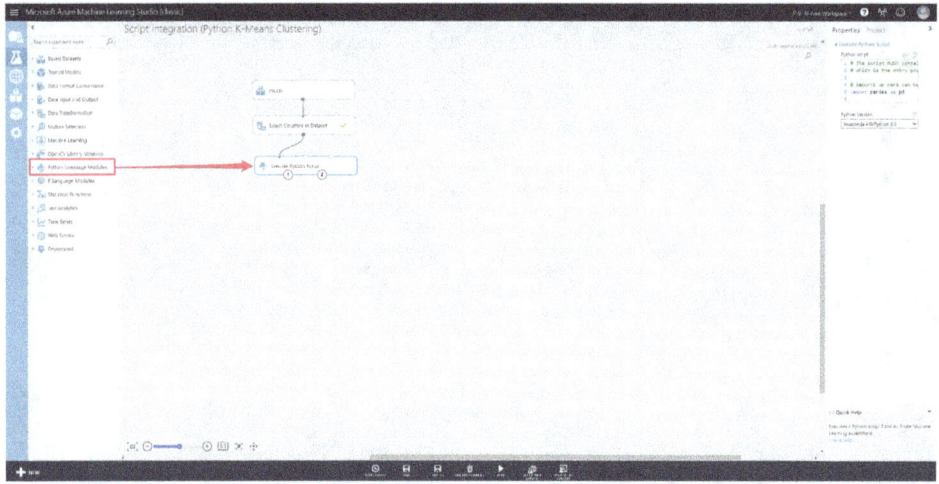

**Figure 8-26**    Arrangement for Python integration

To modify the Python Script, click on the box in the upper right of the Python Script Properties window in Figure 8-27.

To apply the K-Means algorithm, set the values "Sepal.Length" and "Sepal. Width" as features, as shown in Figure 8-28. The code above is written to create three clusters. The code is written below.

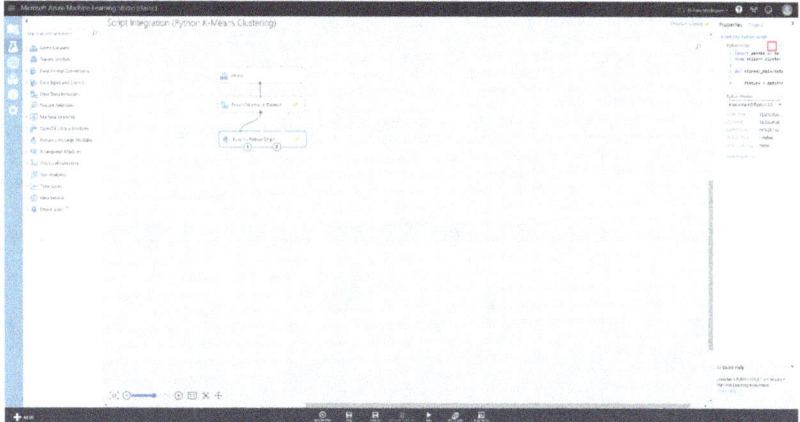

**Figure 8-27**   Python Script modification

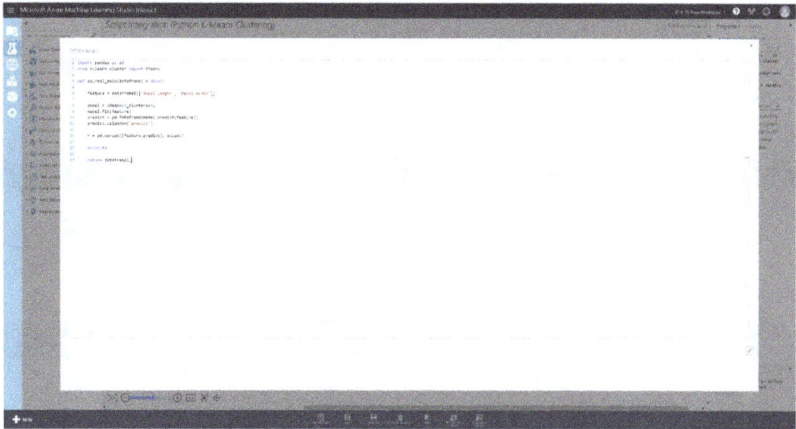

**Figure 8-28**   Python Script editing

## Source (8-28)

```
import pandas as pd
from sklearn.cluster import KMeans

def azureml_main(dataframe1 = None):
        feature = dataframe1[['Sepal.Length', 'Sepal.Width']]

        model = KMeans(n_clusters=3)
        model.fit(feature)
        predict = pd.DataFrame(model.predict(feature))
        predict.columns=['predict']

        r = pd.concat([feature,predict], axis=1)

        print(r)

        return dataframe1,
```

Use the visualization in Figure 8-29 to verify that the results applied to the Python script are intact.

As shown in Figure 8-30, you can see the "Predict" value depending on the values of "Sepal.Length" and "Sepal.Width". Predict values are cluster groups.

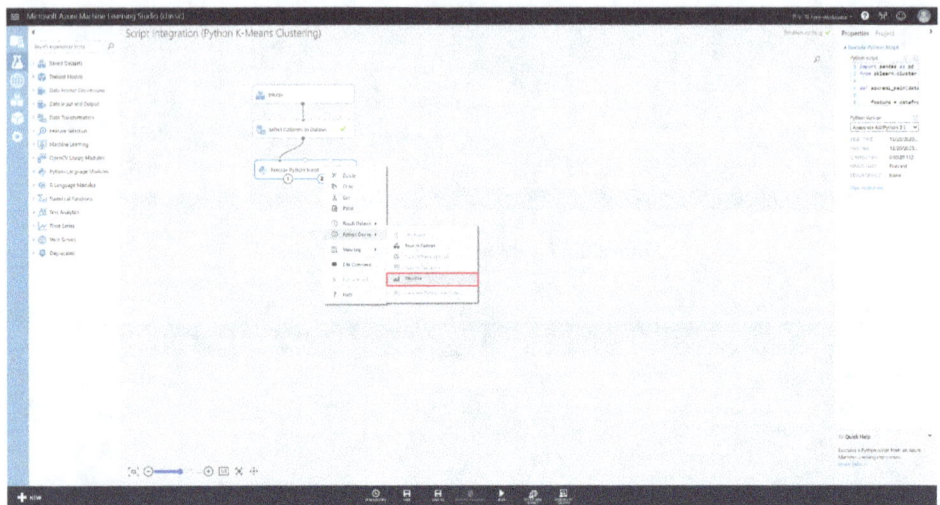

**Figure 8-29**    Visualization of K-Means result

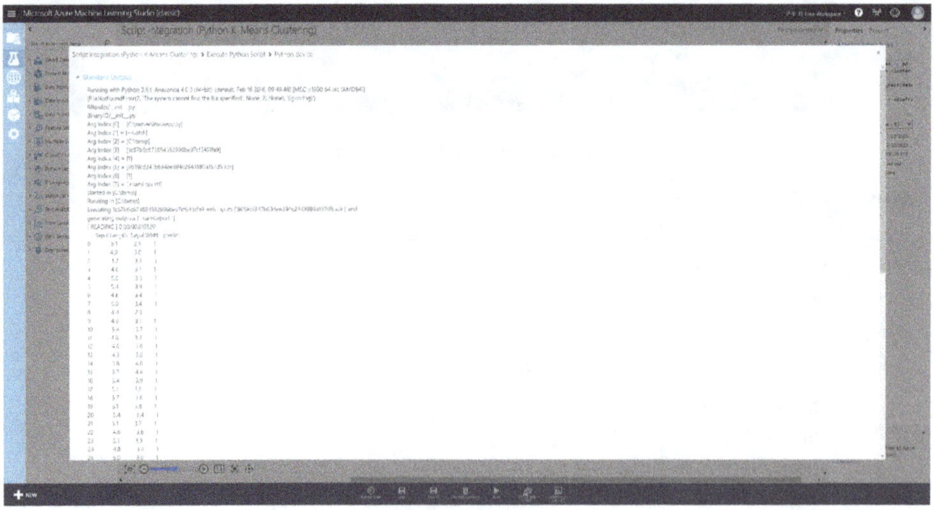

**Figure 8-30**    K-Means result using Python Script (Azure Machine Learning Studio)

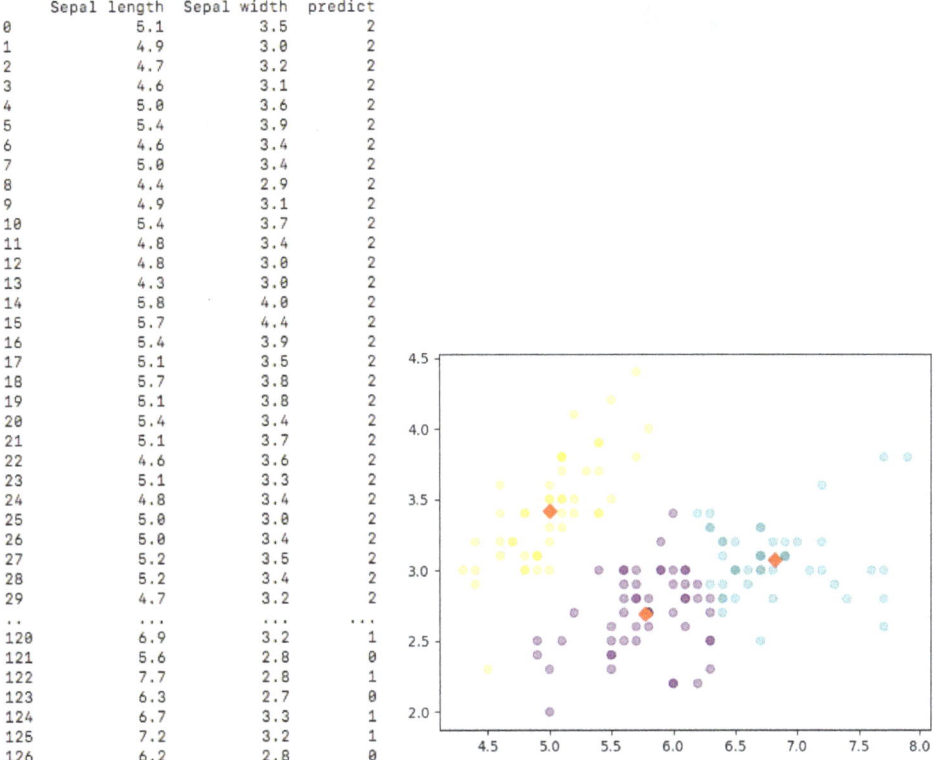

|     | Sepal length | Sepal width | predict |
| --- | --- | --- | --- |
| 0   | 5.1 | 3.5 | 2 |
| 1   | 4.9 | 3.0 | 2 |
| 2   | 4.7 | 3.2 | 2 |
| 3   | 4.6 | 3.1 | 2 |
| 4   | 5.0 | 3.6 | 2 |
| 5   | 5.4 | 3.9 | 2 |
| 6   | 4.6 | 3.4 | 2 |
| 7   | 5.0 | 3.4 | 2 |
| 8   | 4.4 | 2.9 | 2 |
| 9   | 4.9 | 3.1 | 2 |
| 10  | 5.4 | 3.7 | 2 |
| 11  | 4.8 | 3.4 | 2 |
| 12  | 4.8 | 3.0 | 2 |
| 13  | 4.3 | 3.0 | 2 |
| 14  | 5.8 | 4.0 | 2 |
| 15  | 5.7 | 4.4 | 2 |
| 16  | 5.4 | 3.9 | 2 |
| 17  | 5.1 | 3.5 | 2 |
| 18  | 5.7 | 3.8 | 2 |
| 19  | 5.1 | 3.8 | 2 |
| 20  | 5.4 | 3.4 | 2 |
| 21  | 5.1 | 3.7 | 2 |
| 22  | 4.6 | 3.6 | 2 |
| 23  | 5.1 | 3.3 | 2 |
| 24  | 4.8 | 3.4 | 2 |
| 25  | 5.0 | 3.0 | 2 |
| 26  | 5.0 | 3.4 | 2 |
| 27  | 5.2 | 3.5 | 2 |
| 28  | 5.2 | 3.4 | 2 |
| 29  | 4.7 | 3.2 | 2 |
| ..  | ... | ... | ... |
| 120 | 6.9 | 3.2 | 1 |
| 121 | 5.6 | 2.8 | 0 |
| 122 | 7.7 | 2.8 | 1 |
| 123 | 6.3 | 2.7 | 0 |
| 124 | 6.7 | 3.3 | 1 |
| 125 | 7.2 | 3.2 | 1 |
| 126 | 6.2 | 2.8 | 0 |

**Figure 8-31**    K-Means visualization using Python

As shown in Figure 8-31, you can see that the result value applied to Python and the result value applied to Azure Machine Learning Studio are the same.

## 8.3 Practice Questions

Q1. What is the Iris dataset?


Q2. Explain how R and Python scripts improve integration.

# Part II

# Exercises

# Chapter 9

# Exercises

## 9.1 Predicting Car Price Using Regression

Now let's create a predictive model using real data. This chapter uses the linear regression algorithm to predict the price of a car. View the automotive data set at the University of California, Irvine (UCI) Machine Learning Repository. Information on car data sets can be found at the link below. The data set to train on the predictive model can be found by searching in the Azure Machine Learning Studio palette.

https://archive.ics.uci.edu/ml/datasets/Automobile

The number of rows in the data set is 205, with a total of 26 attributes.

The distribution of values in the data consists of the values specified in Table 9-1.

**Table 9-1**   Car dataset

| Attribute Name | Distribution of values |
| --- | --- |
| symbolling | −3, −2, −1, 0, 1, 2, 3 |
| normalized losses | 65~256 |
| make | alfa-romero, audi, bmw, chevrolet, dodge, honda, isuzu, jaguar, mazda, mercedes-benz, mercury, mitsubishi, nissan, peugot, plymouth, porsche, renault, saab, subaru, toyota, volkswagen, volvo |
| fuel-type | diesel, gas |
| aspiration | std, turbo. |
| num-of-doors | four, two. |
| body-style | hardtop, wagon, sedan, hatchback, convertible |
| drive-wheels | 4wd, fwd, rwd. |
| engine-location | front, rear. |
| wheel-base | 86.6~120.9. |
| length | 141.1~208.1 |
| width | 60.3~72.3. |
| height | 47.8~59.8. |
| curb-weight | 1488~4066. |
| engine-type | dohc, dohcv, l, ohc, ohcf, ohcv, rotor. |
| num-of-cylinders | eight, five, four, six, three, twelve, two. |
| engine-size | 61~326. |
| fuel-system | 1bbl, 2bbl, 4bbl, idi, mfi, mpfi, spdi, spfi |
| bore | 2.54~3.94. |
| stroke | 2.07~4.17. |
| compression-ratio | 7~23. |
| horsepower | 48~288. |
| peak-rpm | 4150~6600. |
| city-mpg | 13~49. |
| highway-mpg | 16~54. |
| price | 5118~45400. |

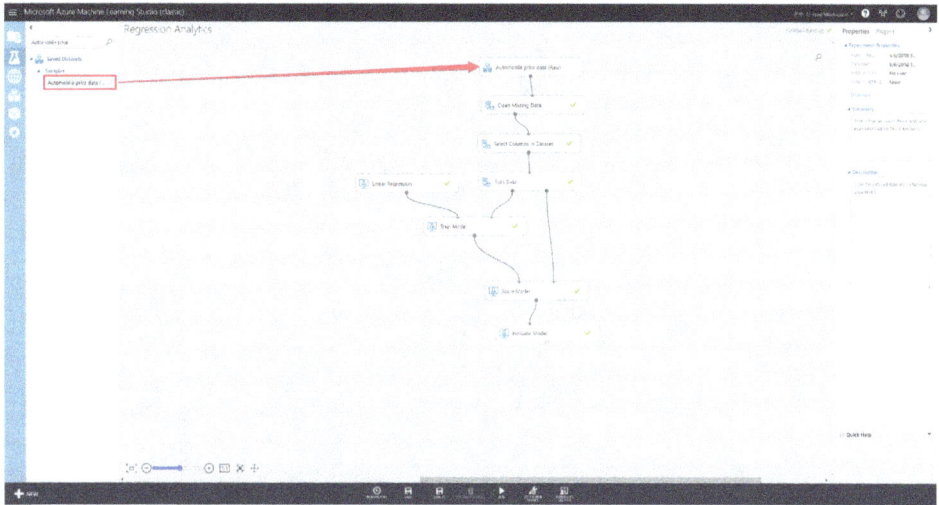

**Figure 9-1** Import sample data

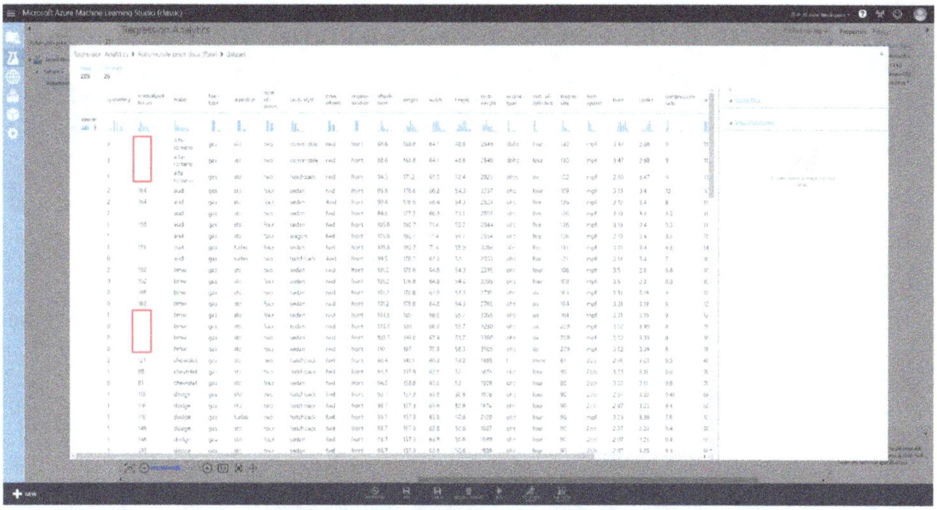

**Figure 9-2** Verifying missing data

Before following the example, import the automobile price data to use in the palette, as shown in Figure 9-1.

However, you can find missing data, as shown in Figure 9-2. The missing data must be processed to ensure accurate learning results.

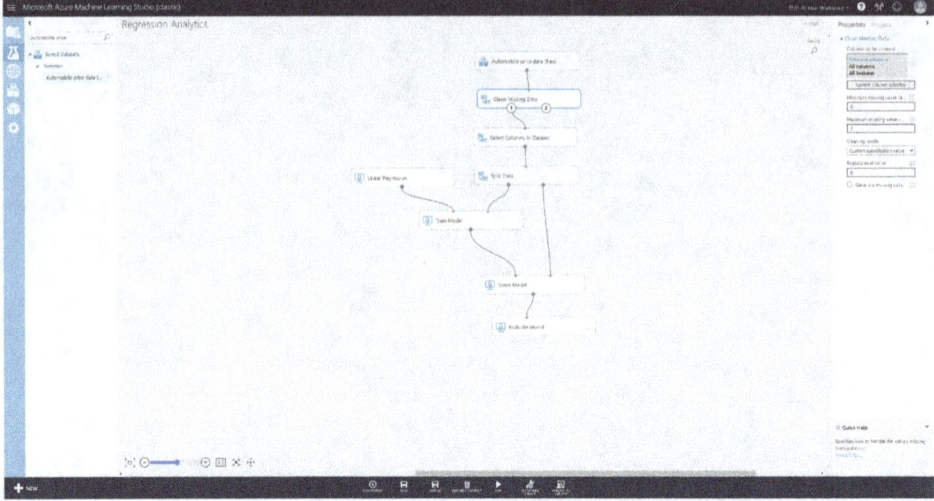

**Figure 9-3**   Clean Missing Data to Preprocess Missing Data

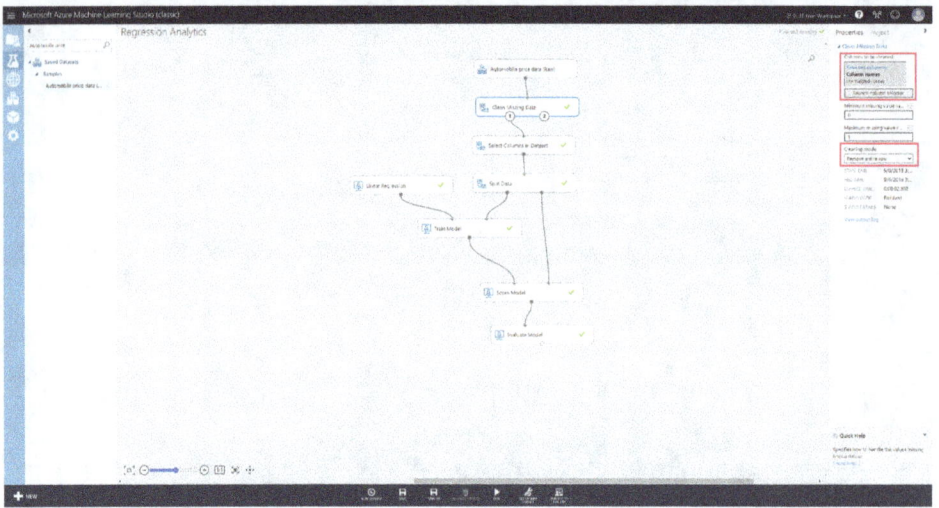

**Figure 9-4**   Clean Missing Data to Preprocess Missing Data

Place a "Clean Missing Data" block on the canvas to fill in the missing data, as shown in Figure 9-3.

Select the missing column, normalized-losses, as shown in Figure 9-4. Then select the option "Remove entire row".

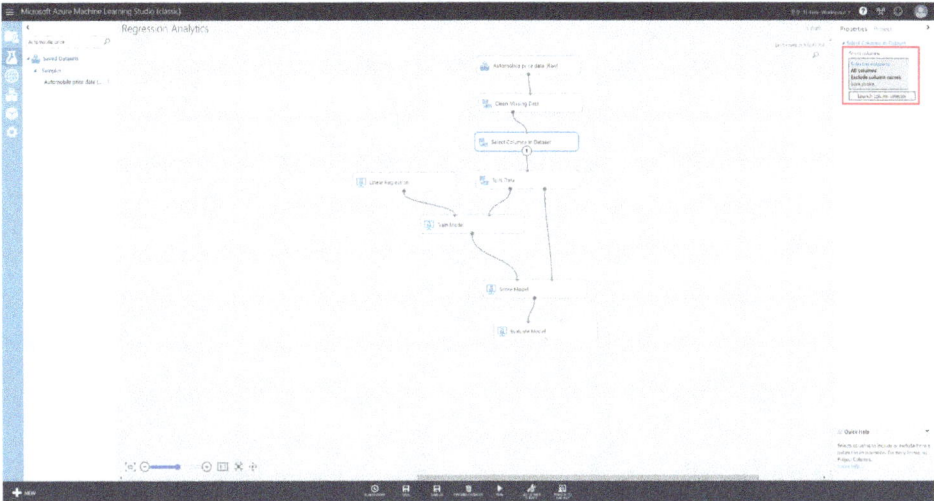

**Figure 9-5**   Select column for training

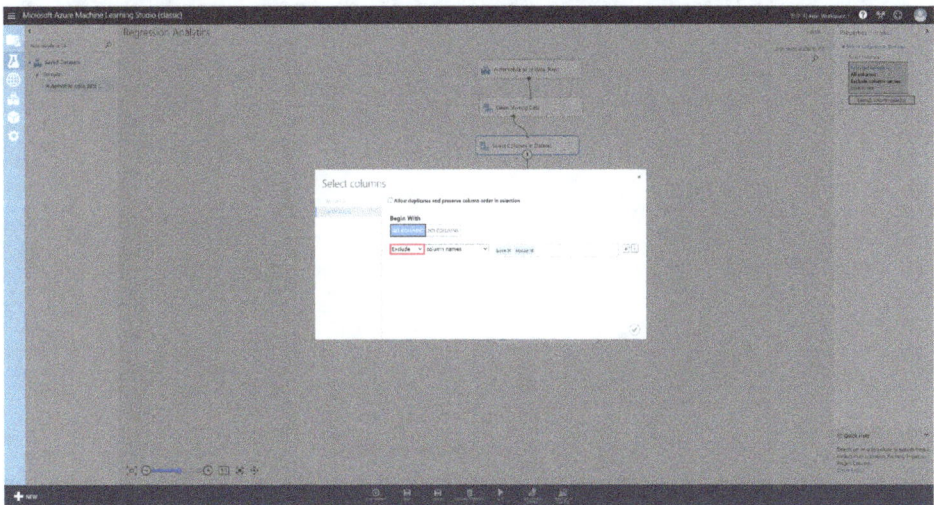

**Figure 9-6**   Select column for training

Before we train, as shown in Figure 9-5, we need to determine which values in the data set to train.

In this example, we will learn to exclude bore and stroke values that are not needed for prediction, as shown in Figure 9-6.

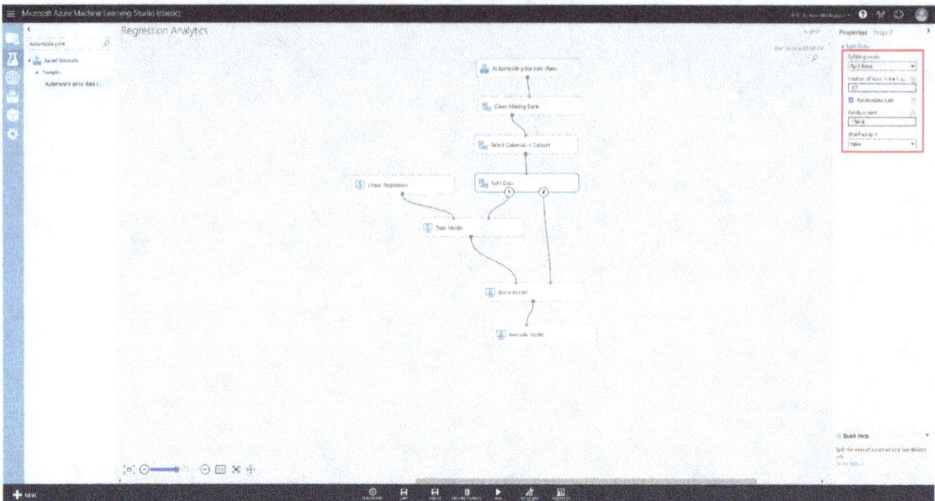

**Figure 9-7**   Data split

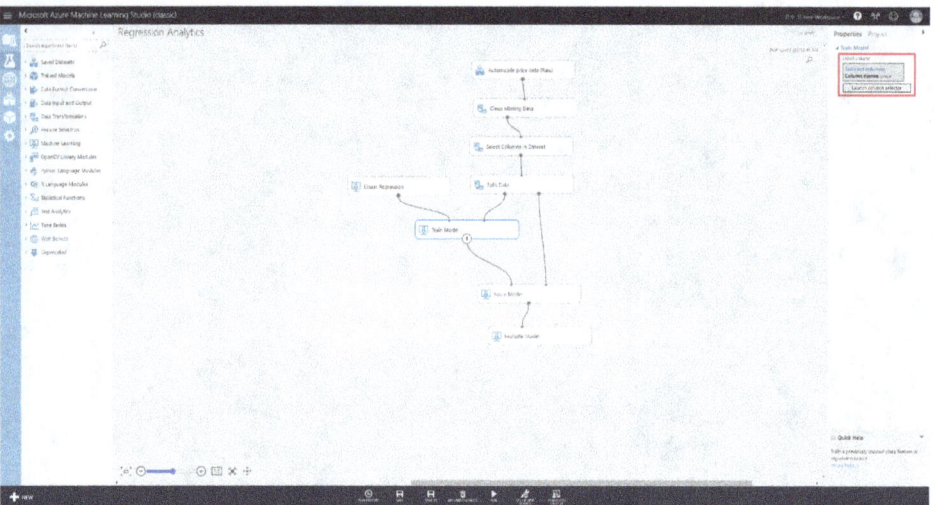

**Figure 9-8**   Choose columns for learning

To avoid overfitting the data, split the data set, as shown in Figure 9-7. The split ratio is 0.7, and the random seed is set to 123.

The algorithm to use is "Linear Regression". Place the "Train Model" block and click "Launch column selector" in the upper right corner of Figure 9-8 to select the columns for training.

Select the column to predict by learning, as shown in Figure 9-9. The project chooses and checks the price value because the goal is to predict the value through the car dataset.

Place the "Score Model" block, as shown in Figure 9-10, and click the "Run" button to learn.

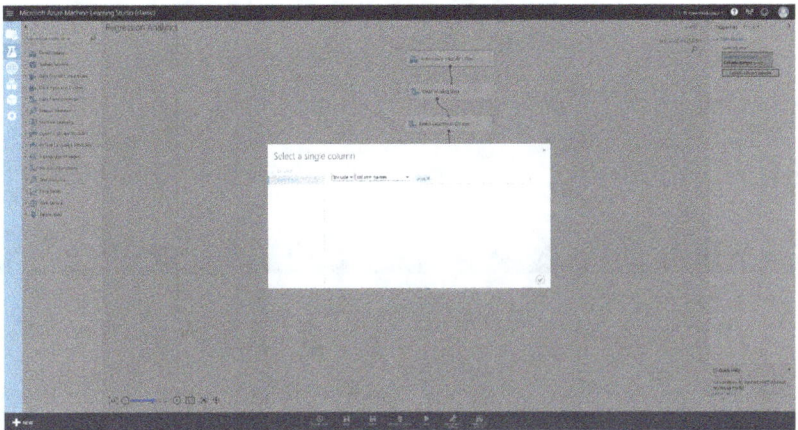

**Figure 9-9**   Choose a value to predict

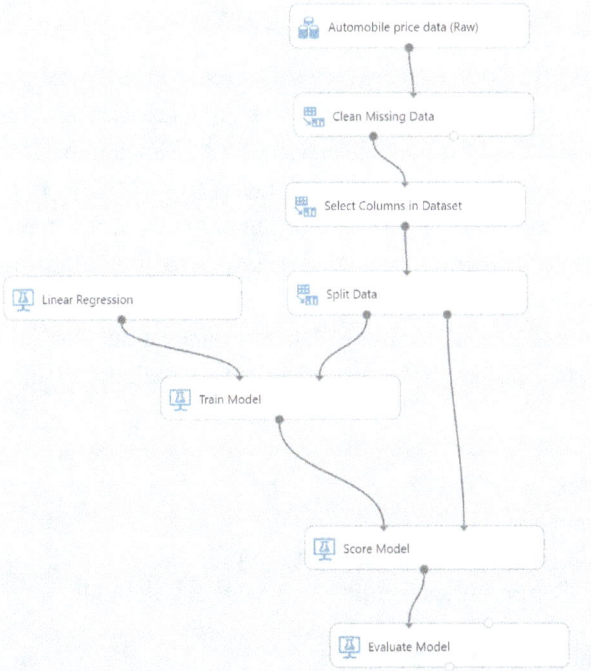

**Figure 9-10**   Arrangement model to view predicted values

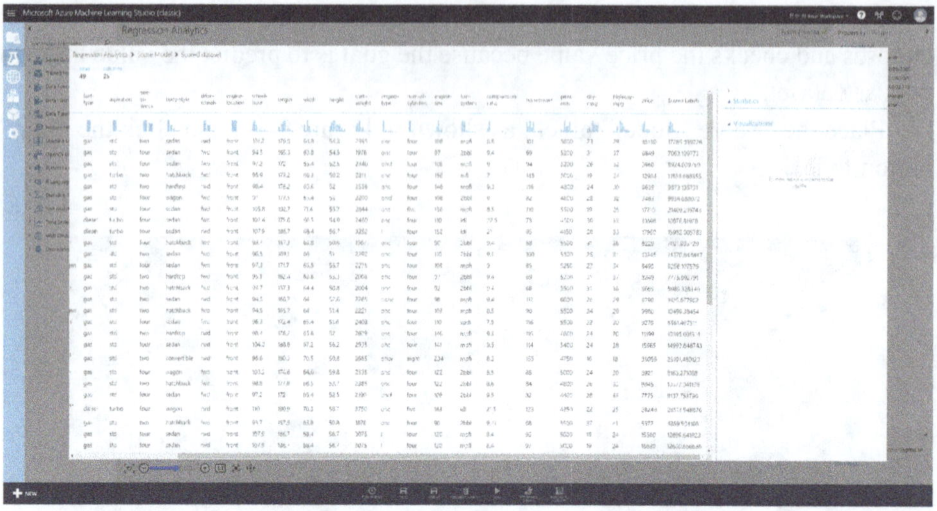

**Figure 9-11**   Prediction result

After learning, right-click on the "Score Model" block to see the visualized predictions. You can see the predicted value in Scored Labels, as shown in Figure 9-11.

## 9.2  Classify News Article Category

This example uses the 2004 Reuters News data set to classify categories. In addition, we will compare the application of the One-vs-All Classifier based on the Multiclass Decision Forest and the Two-class Decision Forest. The parameters of the algorithm were applied by default. This example is taken from Microsoft's AI Gallery. For more information and for details on importing projects, access the links below.

https://gallery.cortanaintelligence.com/Experiment/Multiclass-Classification-News-categorization-2

There are about 23,000 training data sets and about 781,000 test data sets. The existing data set has 103 categories, which belong to the following four categories.

- Corporate-Industrial (CCAT)
- Government and Social (GCAT)
- Economics and Economic Indicators (ECAT)
- Securities and Commodities Trading and Market (MCAT)

The four values are used as labels.

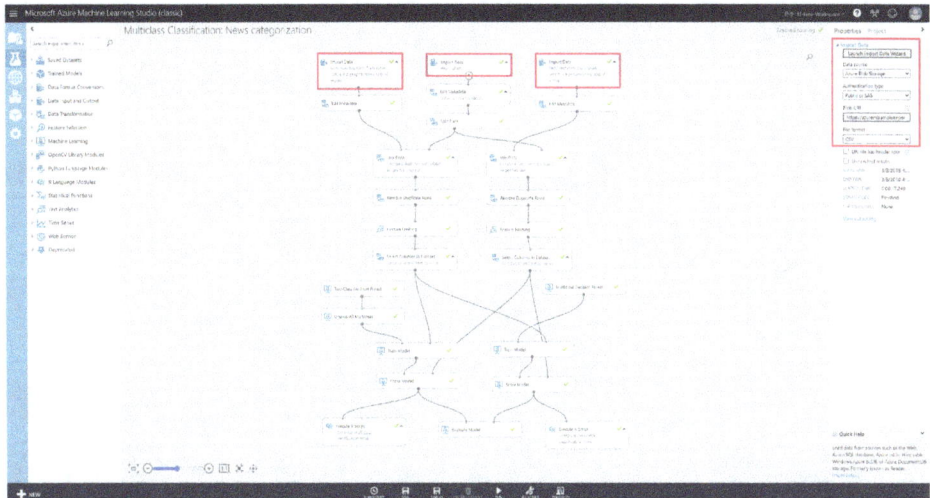

**Figure 9-12**    Sample data through SAS

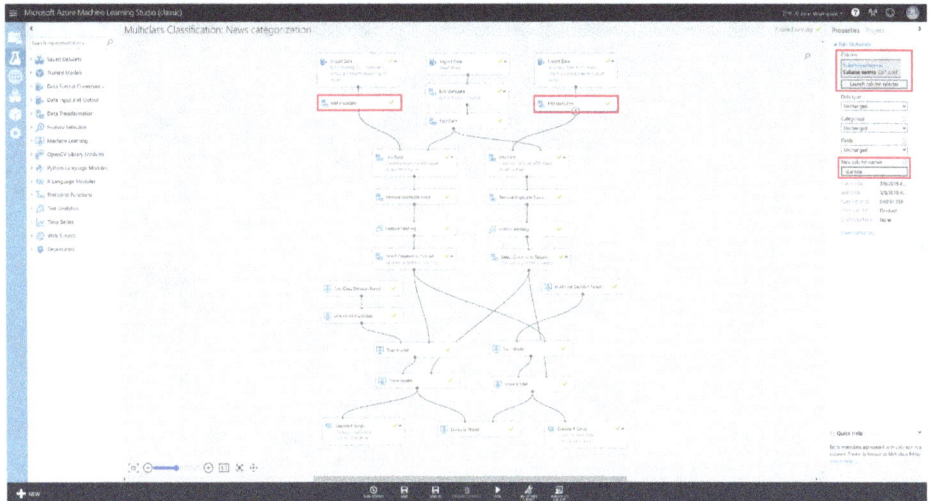

**Figure 9-13**    Set data column name

This project does not provide sample data, but instead pulls data through the functional SAS of the Azure platform, as shown in Figure 9-12. Therefore, import the project using the link above.

Since the sample data imported into SAS is not defined in the column name, change the id and article to Col1 and Col2, respectively, using the Edit Metadata function as shown in Figure 9-13.

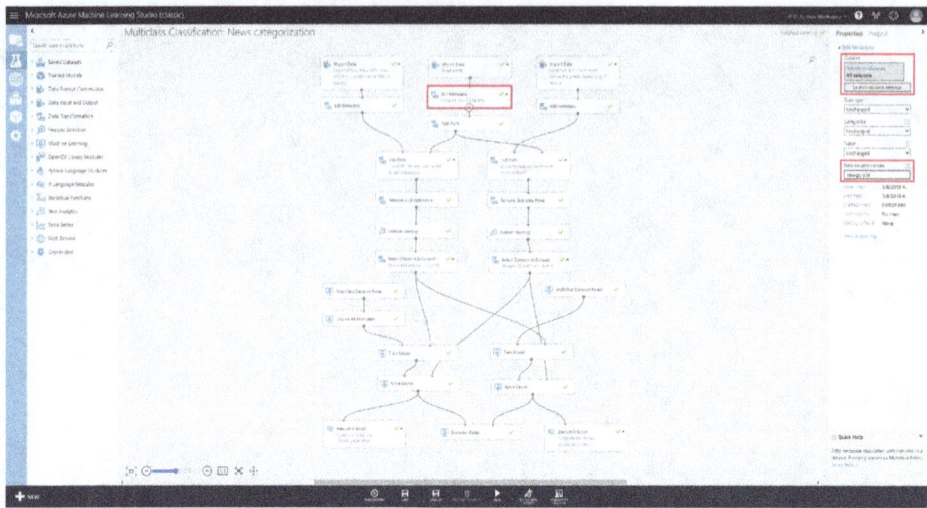

**Figure 9-14** Change the name of the data column

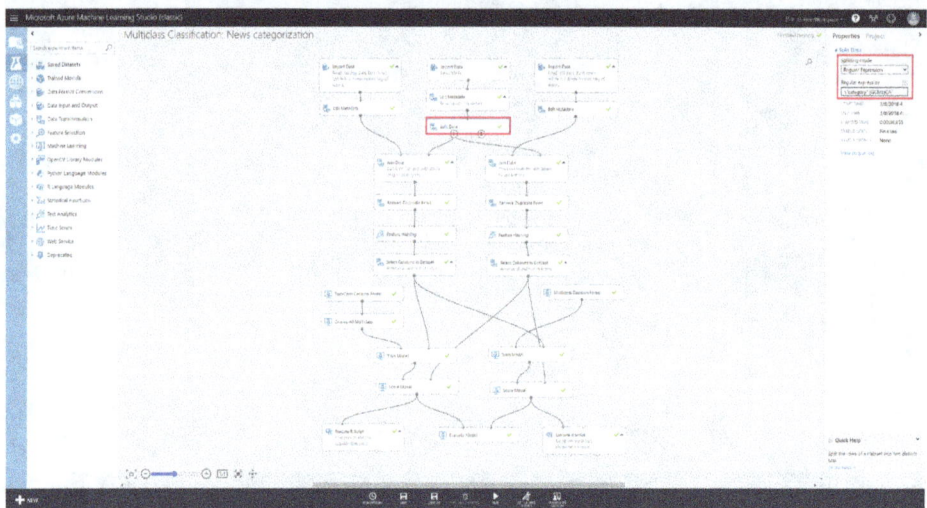

**Figure 9-15** Data split

Similarly, change the datasets in the middle to "category" and "id", as shown in Figure 9-14.

Use the "Split Data" module, as shown in Figure 9-15. Set the Splitting Mode to "Regular Expression".

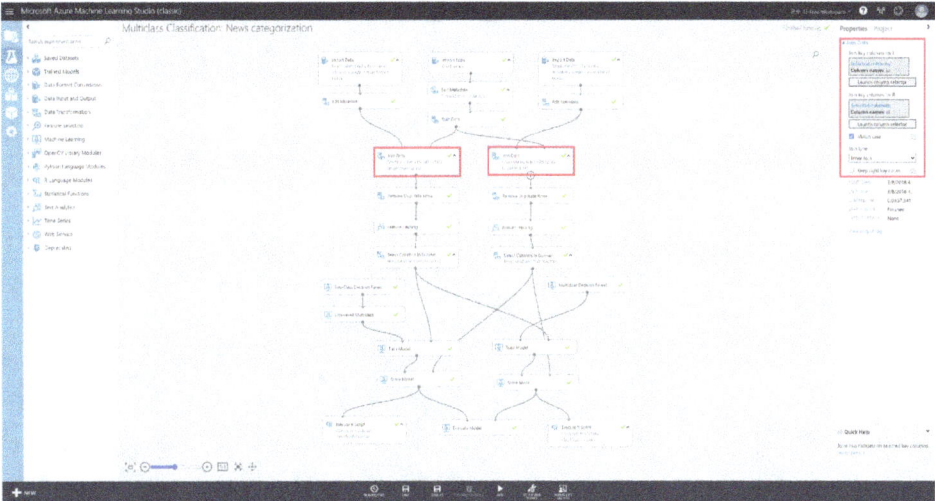

**Figure 9-16** Data join

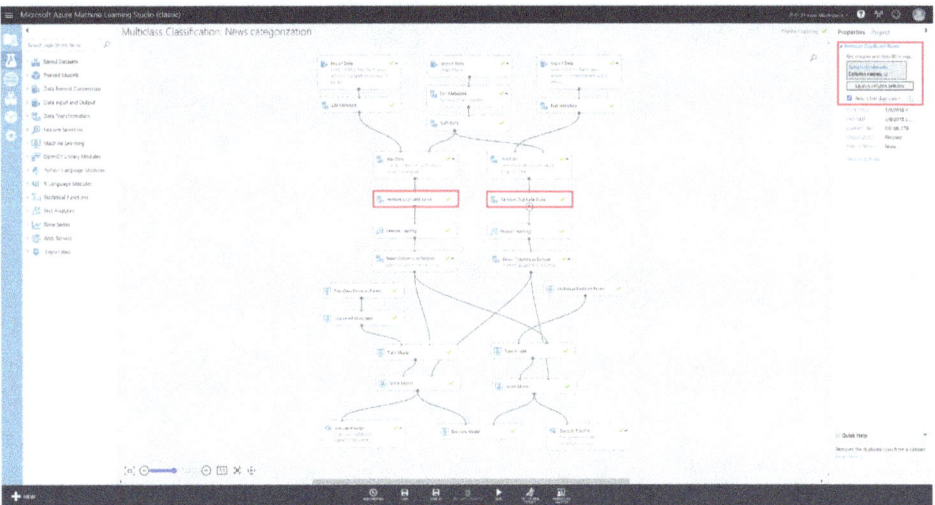

**Figure 9-17** Remove duplicate value

To compare the Multiclass Decision Forest with the One-vs-All Classifier, you need to combine three data sets into two. Join the data based on the "id" column value using the Join Data function, as in Figure 9-16.

To avoid the duplication of combined data sets, remove duplicate id column values using the "Remove Duplicate Rows" feature, as shown in Figure 9-17.

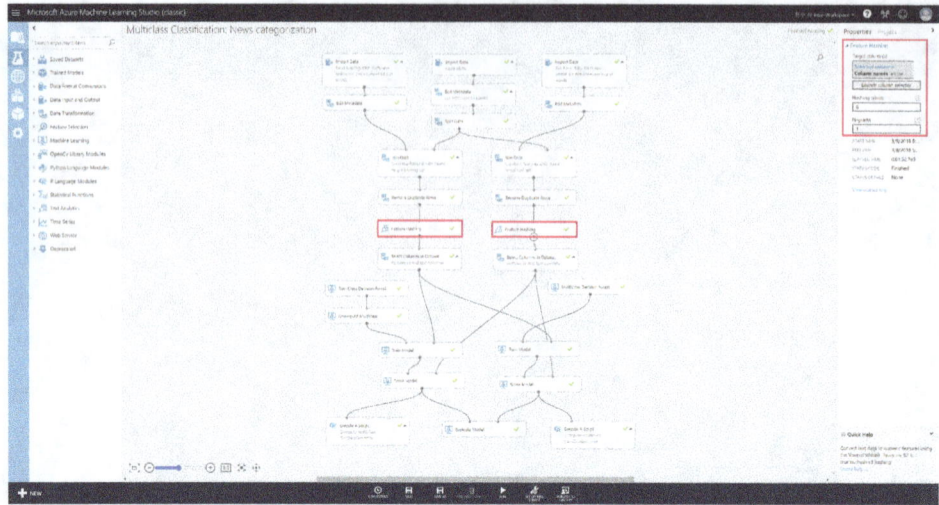

**Figure 9-18**   Set Feature Hashing

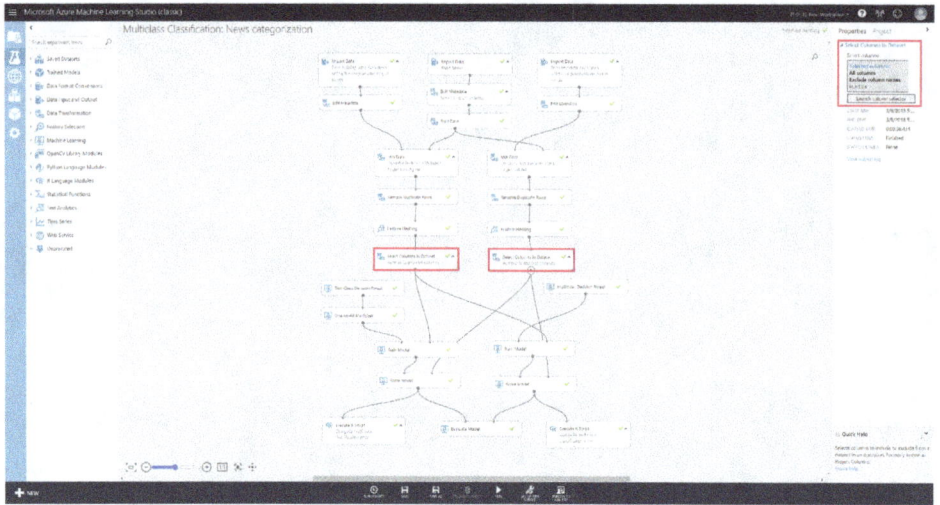

**Figure 9-19**   Column select for training model

Use "Feature Hashing" to convert the plain text value of the article to an integer, as shown in Figure 9-18.

To train the model, select values other than the "id" and "article" columns using the "Select Columns in Dataset" feature, as shown in Figure 9-19.

Click "Launch Column Selector" to enter the excluded value. Figure 9-20 shows the value generated by Feature Hashing.

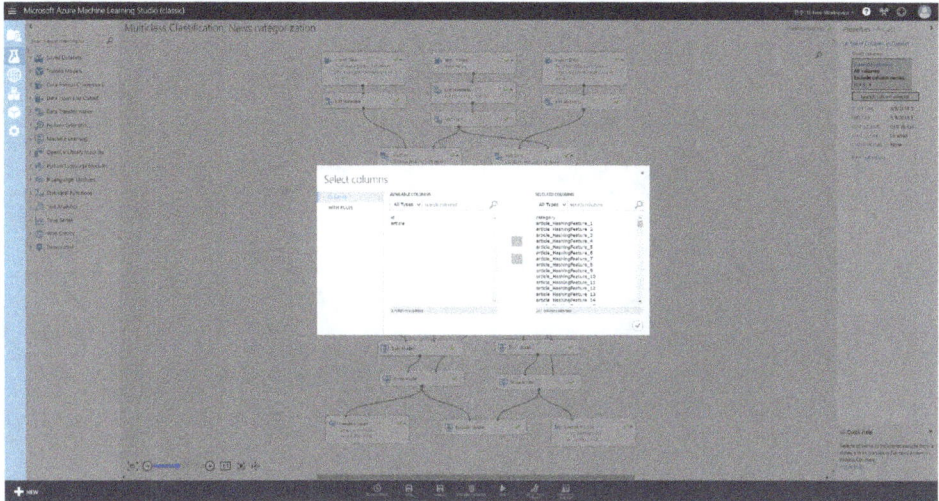

**Figure 9-20**  Value due to Feature Hashing

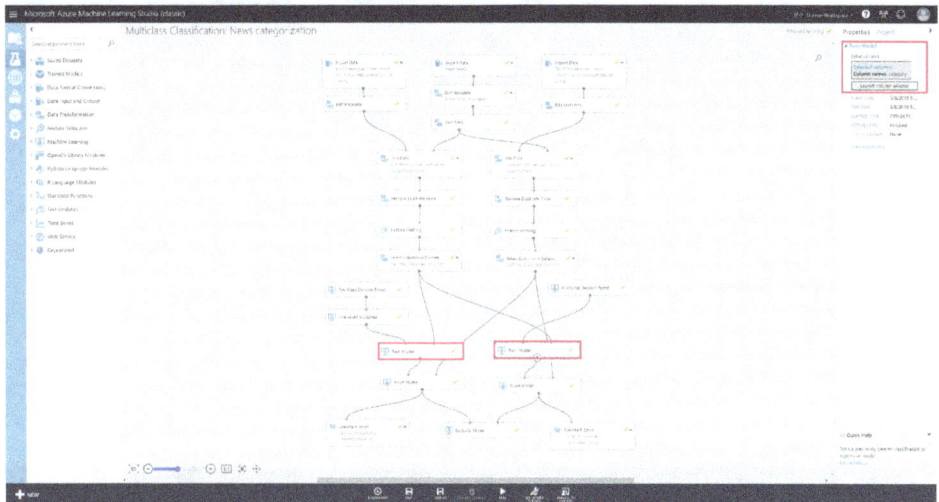

**Figure 9-21**  Train Model Placement for learning

To train the model, arrange the "Train Model" blocks and the algorithm to be applied, as shown in Figure 9-21. The Two-Class Decision Forest and One-vs-All Classifier are arranged on the left side and the Multiclass Decision Forest on the right side. Then select "Launch column selector" to select the values that will result from the model.

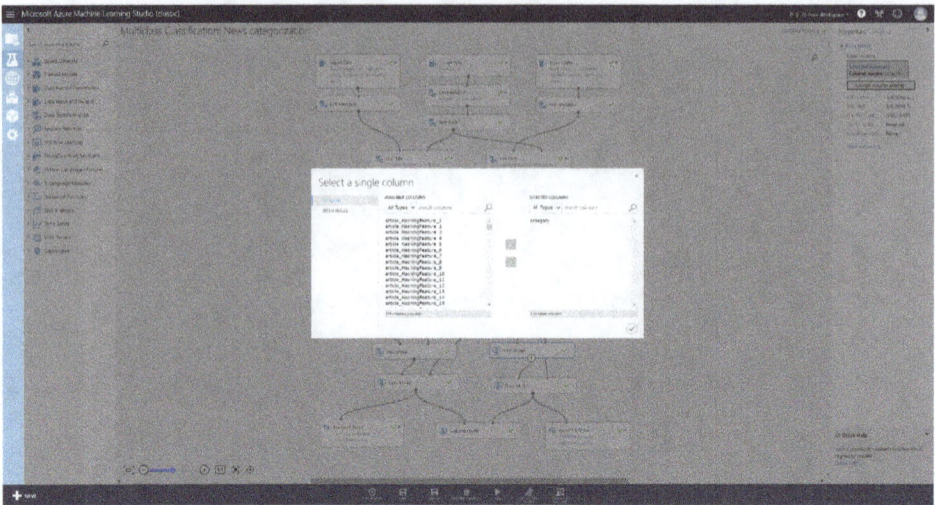

**Figure 9-22**   Label selection

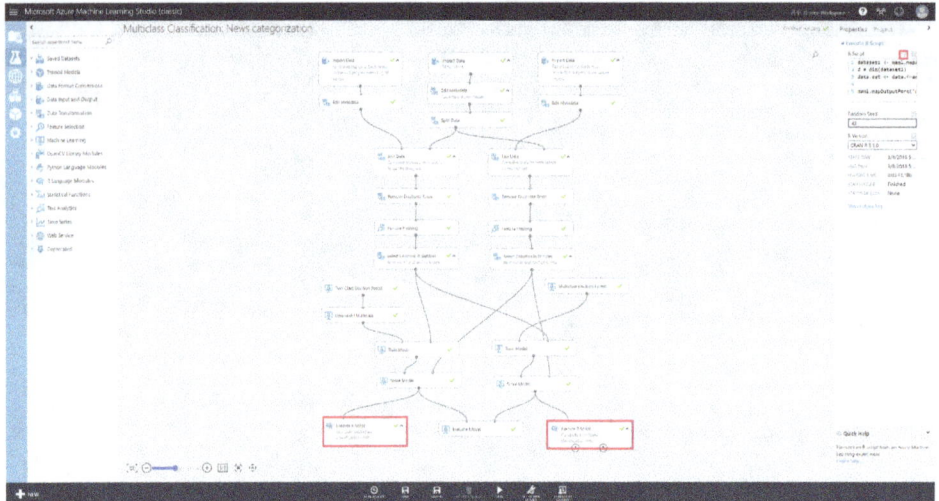

**Figure 9-23**   R Script placement

Select "Launch column selector", and you should see "category" as the result. Select "category" and click the check button to complete, as shown in Figure 9-22.

As shown in Figure 9-23, arrange the "Score Model' block to check the predicted value, and the "Execute R Script" function. Click the "Edit Script" button in the upper right.

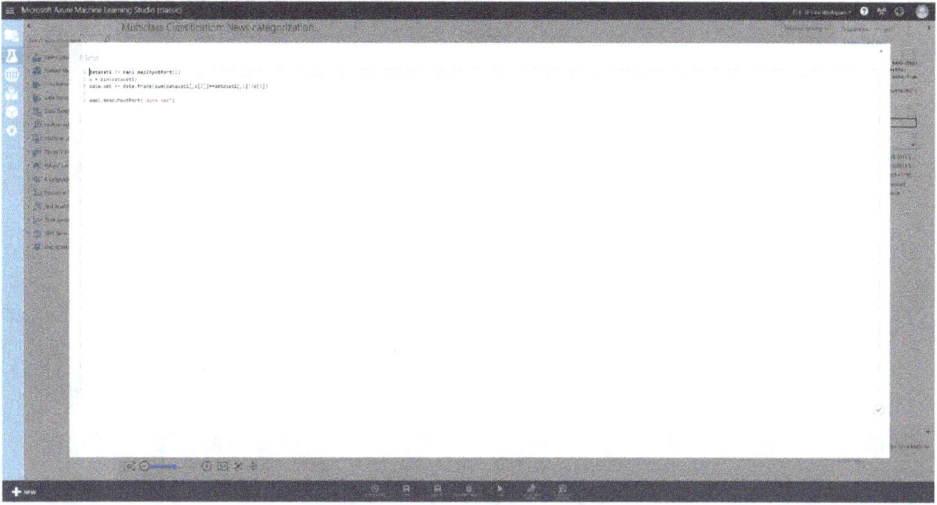

**Figure 9-24**  R Script editing

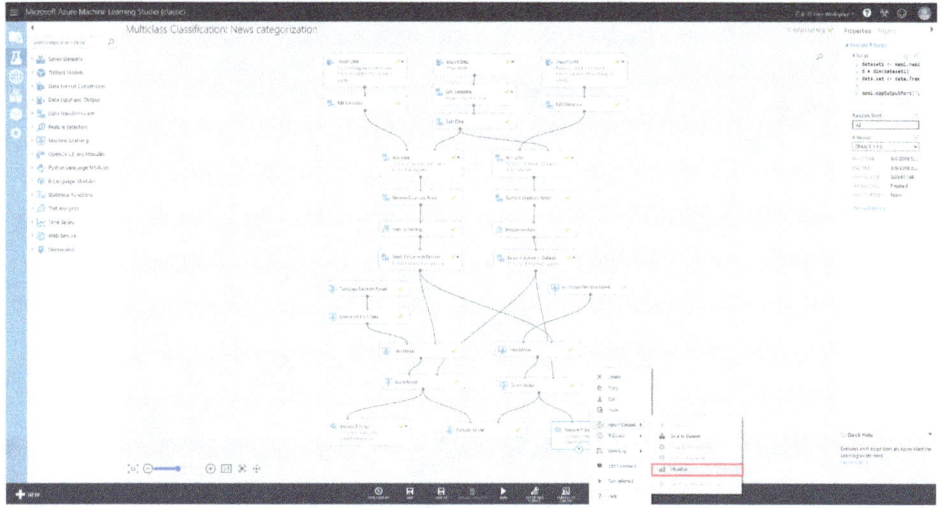

**Figure 9-25**  View R Script results

Write an R script to check for accuracy, as shown in Figure 9-24. The code is written as below.

As shown in Figure 9-25, you can see the result of applying the R script by visualizing the Result Dataset.

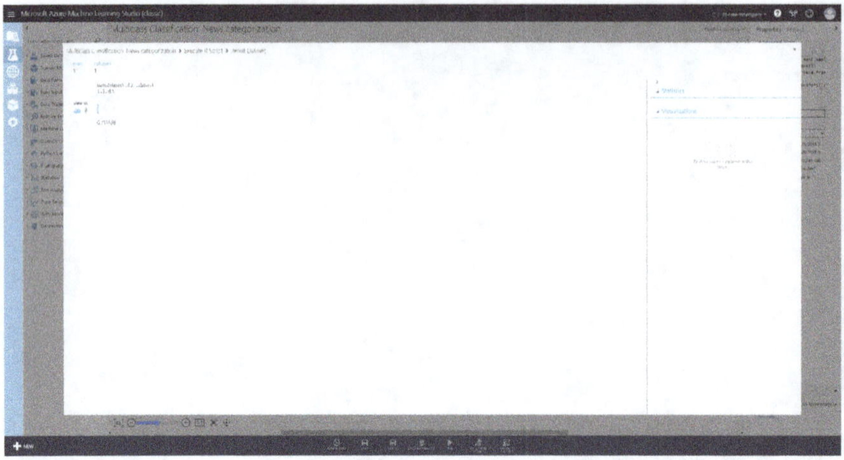

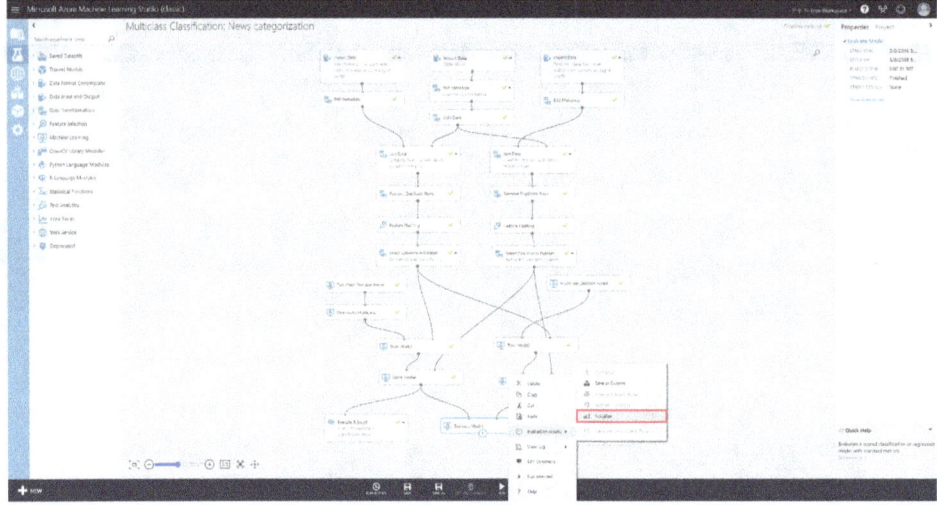

**Figure 9-26**    R Script results

**Figure 9-27**    Evaluate model

As Figure 9-26 shows, the result is accurate.

Place the "Score Model" and "Evaluate Model" blocks as shown in Figure 9-27, click the "Run" button to complete the training, and see the results by visualizing the Evaluation results.

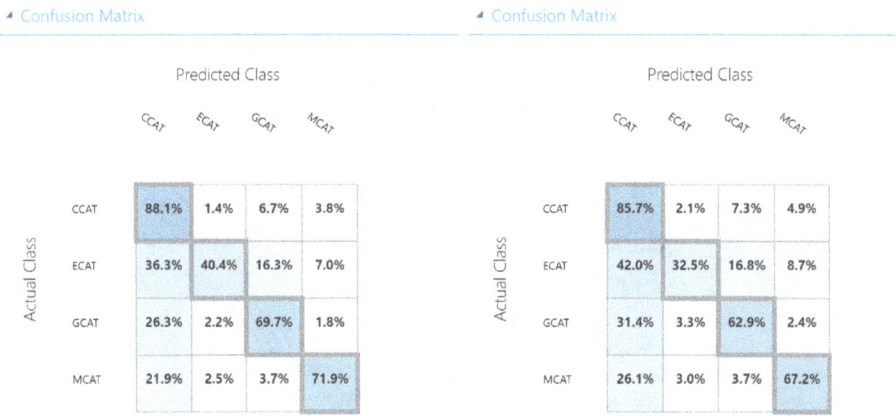

| ◢ Metrics | | | ◢ Metrics | |
| --- | --- | --- | --- | --- |
| Overall accuracy | 0.755351 | | Overall accuracy | 0.711436 |
| Average accuracy | 0.877675 | | Average accuracy | 0.855718 |
| Micro-averaged precision | 0.755351 | | Micro-averaged precision | 0.711436 |
| Macro-averaged precision | 0.756946 | | Macro-averaged precision | 0.697728 |
| Micro-averaged recall | 0.755351 | | Micro-averaged recall | 0.711436 |
| Macro-averaged recall | 0.675049 | | Macro-averaged recall | 0.620682 |

**Figure 9-28**   Evaluation result by algorithm

◢ Confusion Matrix                  ◢ Confusion Matrix

Predicted Class

|  | CCAT | ECAT | GCAT | MCAT |
| --- | --- | --- | --- | --- |
| CCAT | 88.1% | 1.4% | 6.7% | 3.8% |
| ECAT | 36.3% | 40.4% | 16.3% | 7.0% |
| GCAT | 26.3% | 2.2% | 69.7% | 1.8% |
| MCAT | 21.9% | 2.5% | 3.7% | 71.9% |

Predicted Class

|  | CCAT | ECAT | GCAT | MCAT |
| --- | --- | --- | --- | --- |
| CCAT | 85.7% | 2.1% | 7.3% | 4.9% |
| ECAT | 42.0% | 32.5% | 16.8% | 8.7% |
| GCAT | 31.4% | 3.3% | 62.9% | 2.4% |
| MCAT | 26.1% | 3.0% | 3.7% | 67.2% |

**Figure 9-29**   Visualize prediction results by algorithm

Figure 9-28 shows that the accuracy of the One-vs-All Classifier is 0.75, and the accuracy of the Multiclass Decision Forest is 0.71.

Figure 9-29 provides an intuitive view of the visualized predictions of the One-vs-All Classifier and the Multiclass Decision Forest. From the figure, we can see that CCAT is good at predicting, but prediction is difficult in the case of ECAT.

## 9.3 Exploring Credit Risk Groups Using Anomaly Detection

This practice uses the Statlog data set provided by UCI. We will apply and compare the anomaly detection algorithms PCA-Based Anomaly Detection and One-Class Support Vector Machine to the data set. Information from the Statlog data set can be found at the University of California, Irvine (UCI) Machine Learning Repository. Check out the link below.

http://archive.ics.uci.edu/ml/datasets/Statlog+%28German+Credit+Data%29

Table 9-2 shows configuration of German credit data set.

**Table 9-2**   Stalog (German Credit Data) data set

| Attribute name | Distribution of values |
| --- | --- |
| Col1 | Status of existing checking account<br>A11: ... < 0 DM<br>A12: 0 <= ... < 200 DM<br>A13: ... >= 200 DM/salary assignments for at least 1 year<br>A14: no checking account |
| Col2 | Duration in month (numerical) |
| Col3 | Credit history<br>A30: no credits taken/ all credits paid back duly<br>A31: all credits at this bank paid back duly<br>A32: existing credits paid back duly till now<br>A33: delay in paying off in the past<br>A34: critical account/ other credits existing (not at this bank) |
| Col4 | Purpose<br>A40: car (new)<br>A41: car (used)<br>A42: furniture/equipment<br>A43: radio/television<br>A44: domestic appliances<br>A45: repairs<br>A46: education<br>A47: (vacation — does not exist?)<br>A48: retraining<br>A49: business<br>A410: others |
| Col5 | Credit amount (numerical) |
| Col6 | Savings account/bonds<br>A61: ... < 100 DM<br>A62: 100 <= ... < 500 DM<br>A63: 500 <= ... < 1000 DM<br>A64: ... >= 1000 DM<br>A65: unknown/ no savings account |
| Col7 | Present employment since<br>A71: unemployed<br>A72: ... < 1 year<br>A73: 1 <= ... < 4 years<br>A74: 4 <= ... < 7 years<br>A75: ... >= 7 years |
| Col8 | Installment rate in percentage of disposable income (numerical) |
| Col9 | Personal status and sex<br>A91: male: divorced/separated |

**Table 9-2** (*Continued*)

| Attribute name | Distribution of values |
|---|---|
| | A92: female: divorced/separated/married |
| | A93: male: single |
| | A94: male: married/widowed |
| | A95: female: single |
| Col10 | Other debtors/guarantors |
| | A101: none |
| | A102: co-applicant |
| | A103: guarantor |
| Col11 | Present residence since (numerical) |
| Col12 | Property |
| | A121: real estate |
| | A122: if not A121: building society savings agreement/ life insurance |
| | A123: if not A121/A122: car or other, not in attribute 6 |
| | A124: unknown/no property |
| Col13 | Age in years (numerical) |
| Col14 | Other installment plans |
| | A141: bank |
| | A142: stores |
| | A143: none |
| Col15 | Housing |
| | A151: rent |
| | A152: own |
| | A153: for free |
| Col16 | Number of existing credits at this bank (numerical) |
| Col17 | Job |
| | A171: unemployed/ unskilled — non-resident |
| | A172: unskilled — resident |
| | A173: skilled employee/official |
| | A174: management/ self-employed/highly qualified employee/ officer |
| Col18 | Number of people being liable to provide maintenance for (numerical) |
| Col19 | Telephone |
| | A191: none |
| | A192: yes, registered under the customer's name |
| Col20 | foreign worker |
| | A201: yes |
| | A202: no |
| Col21 | Label |
| | 1: Normal |
| | 2: Risky |

It is examples from Microsoft's AI Gallery. For more information and details on importing the project, access the following links. https://gallery.azure.ai/ Experiment/Anomaly-Detection-Credit-Risk-5

For clearer workflow, change the column name from "Col21" to "Label", as shown in Figure 9-30.

As shown in Figure 9-31, the total data set is divided into 75% training data and 25% test data.

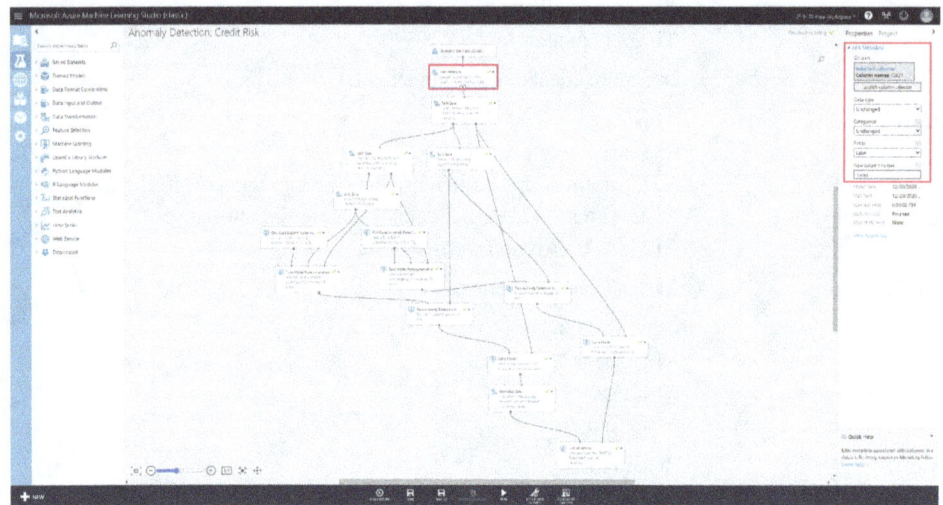

**Figure 9-30**   Set Edit Metadatat

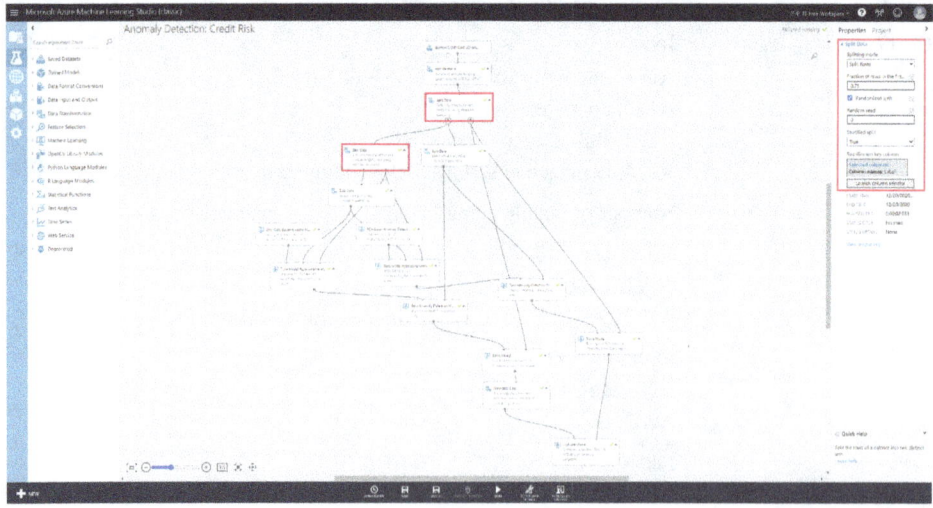

**Figure 9-31**   Data split

Anomaly detection learns data that contains a single class (assuming Normal). Remove the row containing the "Risky" label using a regular expression filter, as shown in Figure 9-32.

Retrieve the "One-Class Support Vector Machine" block from the palette, set the Trainer mode to "Parameter Range", and set the parameter values as shown in Figure 9-33.

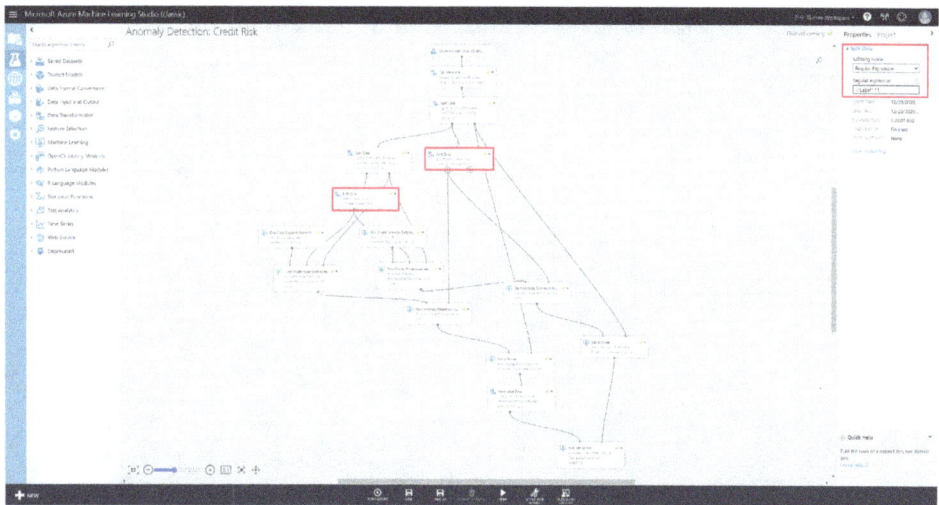

**Figure 9-32** Remove unnecessary rows

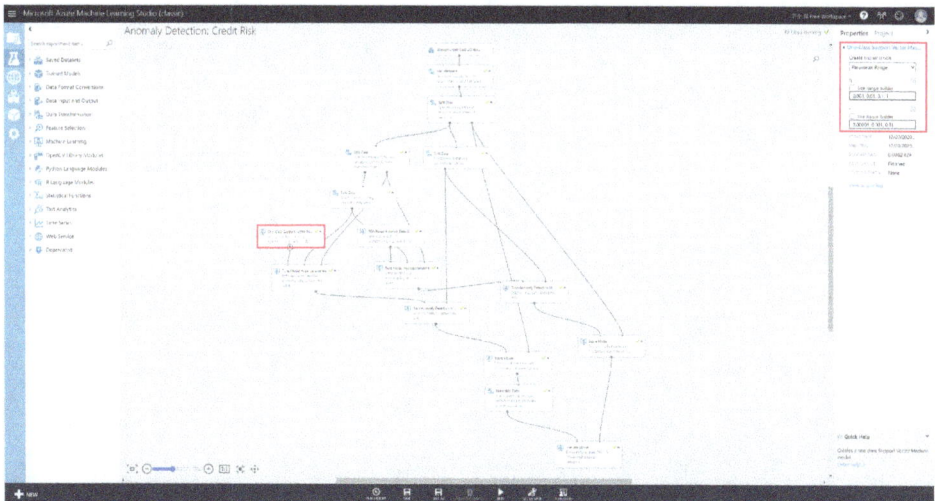

**Figure 9-33** Set One-Class Support Vector Machine parameter

Retrieve the "PCA-Based Anomaly Detection" block from the palette, set Trainer mode to "Parameter Range", and set the parameter values as shown in Figure 9-34.

Retrieve the "Tune Model Hyperparameters" block from the palette and set the "Specify Parameter Sweeping" value to "Entire grid", as shown in Figure 9-35.

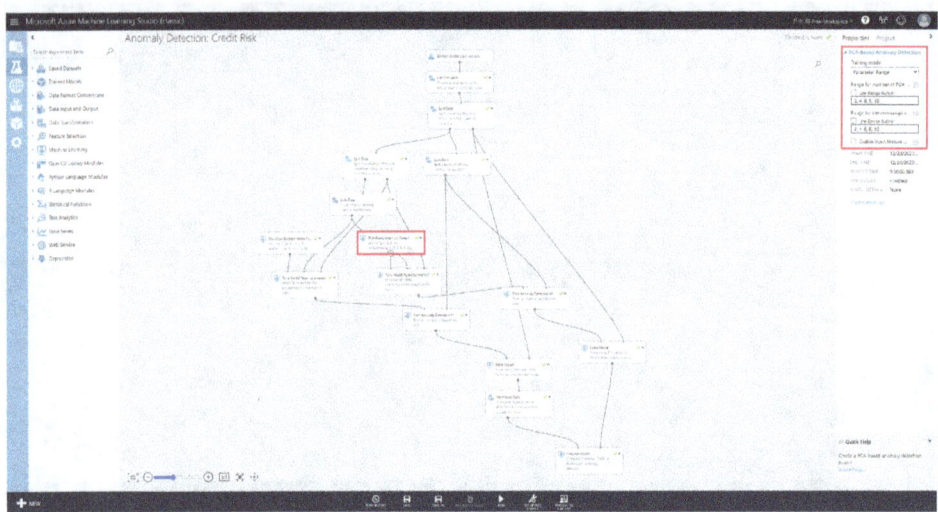

**Figure 9-34**　Set PCA-Based Anomaly Detection parameter

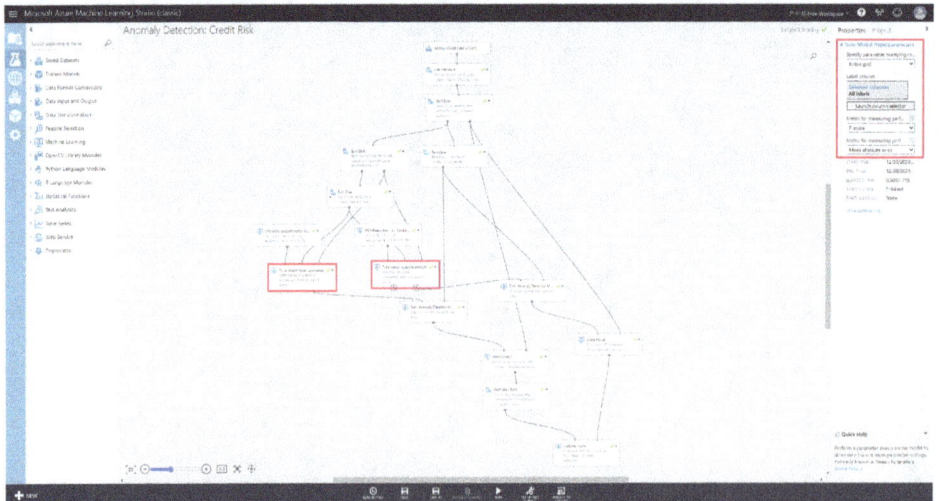

**Figure 9-35**　Tune Model Hyperparameters parameter setting

Retrieve the "Training Model" block from the palette and connect it, as shown in Figure 9-36.

Retrieve the "Score Model" block from the palette, connect it as shown in Figure 9-37, and check the "Append score columns" to output.

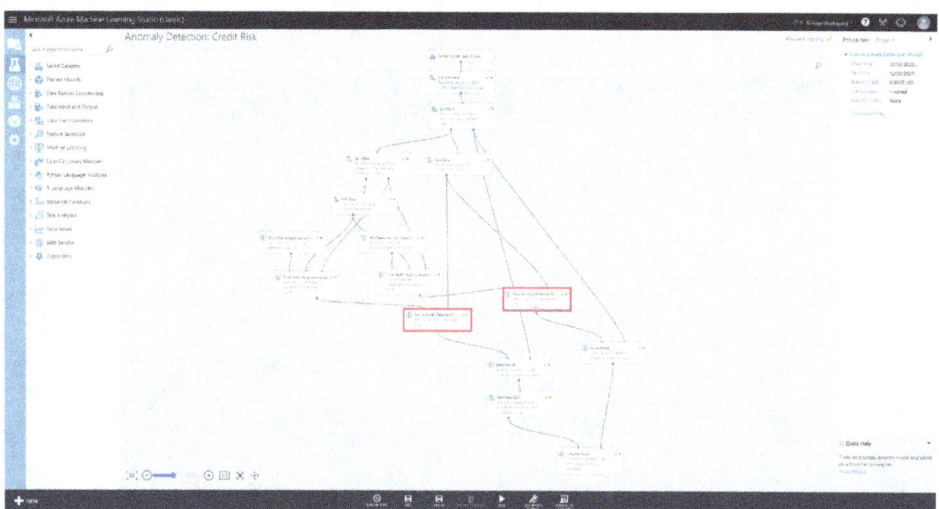

**Figure 9-36** Training Model placement for learning

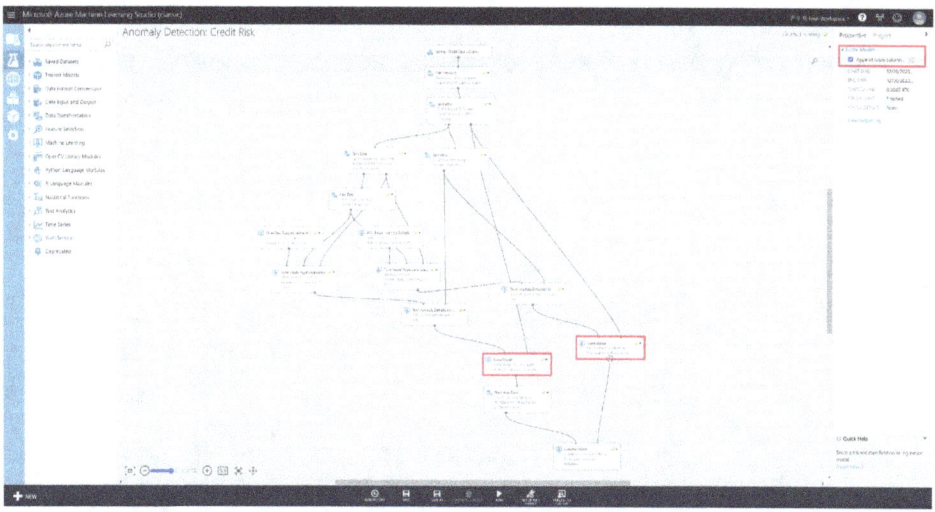

**Figure 9-37** Score Model placement for prediction

The probability of the predicted value is negative, as shown in Figure 9-38. Since probability cannot be a negative value, it must be corrected.

Therefore, it is necessary to correct the negative probability value using the Normalize Data function. Retrieve the "Normalize Data" block from the palette and connect it as shown in Figure 9-39. Then, change the Transformation method to "Logistic" and select the column "Scored Probabilities".

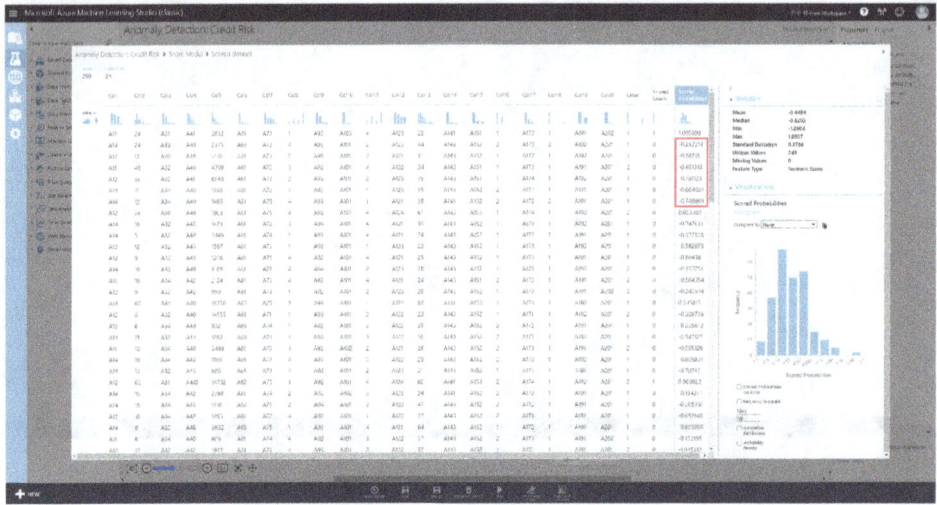

**Figure 9-38**   Specific Score model

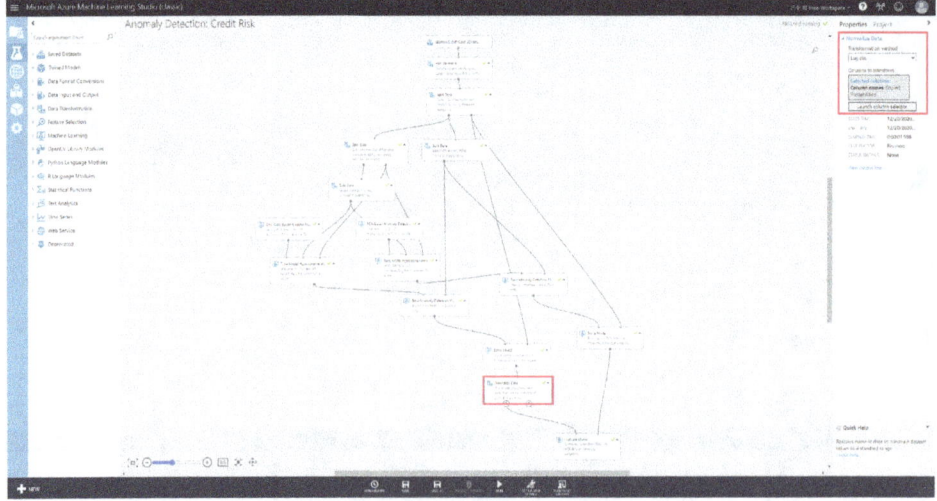

**Figure 9-39**   Normalize data

As shown in Figure 9-40, the negative probability value has changed to a positive value.

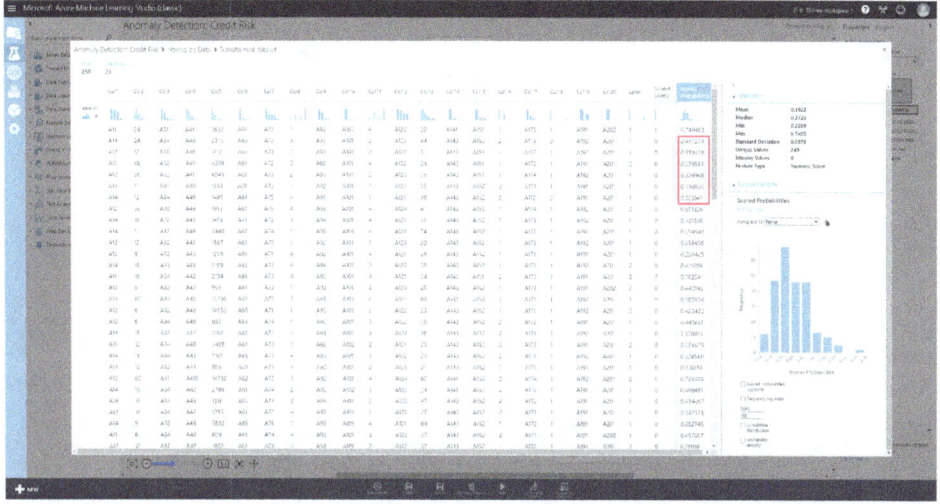

**Figure 9-40**   Normalized value

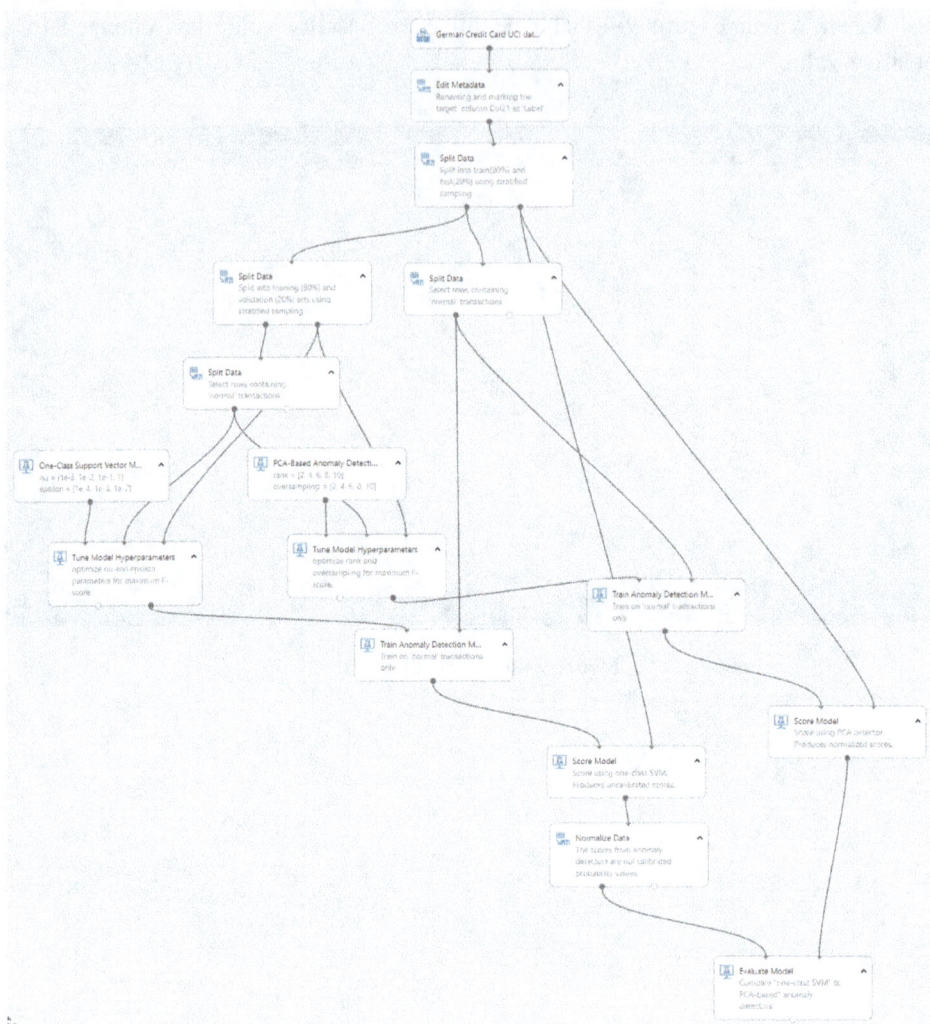

**Figure 9-41**   Model arrangement

Figure 9-41 shows the fault anomaly model arrangement. Click the "Run" button to learn and see the predictions.

ROC PRECISION/RECALL LIFT

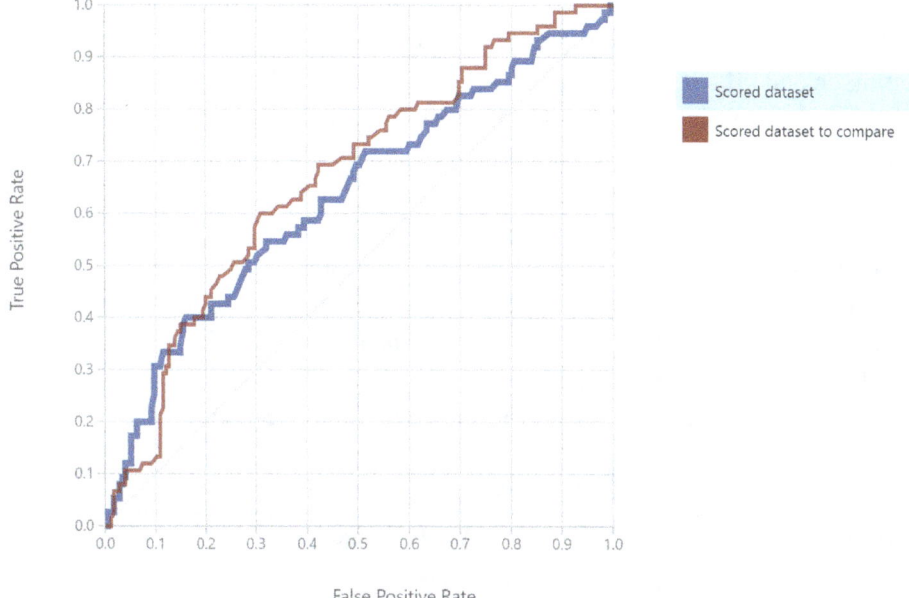

**Figure 9-42** ROC of the trained model

| True Positive | False Negative | Accuracy | Precision | Threshold | | AUC |
|---|---|---|---|---|---|---|
| 15 | 60 | 0.716 | 0.577 | 0.5 | | 0.633 |

| False Positive | True Negative | Recall | F1 Score |
|---|---|---|---|
| 11 | 164 | 0.200 | 0.297 |

| Positive Label | Negative Label |
|---|---|
| 2 | 1 |

| Score Bin | Positive Examples | Negative Examples | Fraction Above Threshold | Accuracy | F1 Score | Precision | Recall | Negative Precision | Negative Recall | Cumulative AUC |
|---|---|---|---|---|---|---|---|---|---|---|
| (0.900,1.000] | 0 | 0 | 0.000 | 0.700 | 0.000 | 1.000 | 0.000 | 0.700 | 1.000 | 0.000 |
| (0.800,0.900] | 0 | 0 | 0.000 | 0.700 | 0.000 | 1.000 | 0.000 | 0.700 | 1.000 | 0.000 |
| (0.700,0.800] | 1 | 1 | 0.008 | 0.700 | 0.026 | 0.500 | 0.013 | 0.702 | 0.994 | 0.000 |
| (0.600,0.700] | 2 | 2 | 0.024 | 0.700 | 0.074 | 0.500 | 0.040 | 0.705 | 0.983 | 0.000 |
| (0.500,0.600] | 12 | 8 | 0.104 | 0.716 | 0.297 | 0.577 | 0.200 | 0.732 | 0.937 | 0.005 |
| (0.400,0.500] | 27 | 54 | 0.428 | 0.608 | 0.462 | 0.393 | 0.560 | 0.769 | 0.629 | 0.133 |
| (0.300,0.400] | 28 | 85 | 0.880 | 0.380 | 0.475 | 0.318 | 0.933 | 0.833 | 0.143 | 0.497 |
| (0.200,0.300] | 5 | 25 | 1.000 | 0.300 | 0.462 | 0.300 | 1.000 | 1.000 | 0.000 | 0.633 |
| (0.100,0.200] | 0 | 0 | 1.000 | 0.300 | 0.462 | 0.300 | 1.000 | 1.000 | 0.000 | 0.633 |
| (0.000,0.100] | 0 | 0 | 1.000 | 0.300 | 0.462 | 0.300 | 1.000 | 1.000 | 0.000 | 0.633 |

**Figure 9-43** Evaluation results

Figure 9-42 shows the receiver operating characteristic curve (ROC) of the trained model. The ROC is a graph showing the ratio of true positives to false positives. This allows us to evaluate the performance of the trained model.

Figure 9-43 shows the evaluation results of the prediction model.

## 9.4 Predicting the Number of People Getting on and Off at Gangnam Station in the Morning Rush Hours

In this chapter, we will apply linear regression to predict the number of people getting on and off at Gangnam station in the morning rush hours using the statistical data of people getting on and off Line 1~4 (Seoul Metro) provided by the public data portal.

Please access the website address below for data of the number of passengers getting on and off the subway lines 1~4.

https://www.data.go.kr/dataset/15003169/fileData.do

This data is of people getting on and off at Gangnam station between Jan. 1. 2016 to Oct. 31.2016; the number of rows in the data set is 610, and the attributes are different for each provided data set.

The composition of the data set is shown in Table 9-3 below.

**Table 9-3**   Getting on and off data set

| Attribute | Attribute description |
|-----------|----------------------|
| Date | The date passengers got on and off |
| Station | The name of the station where passengers got on and off |
| Line | The line of the station where passengers got on and off |
| Type | The type of the getting on or off |
| Discount | The distribution of passengers such as general, child of school-going age, senior (January-May 2016) |
| Time | The time passengers got on and off (unit = an hour) |

Let's take a look at how important attributes are configured before applying.

Before applying the data to Azure Machine Learning Studio, preprocessing is needed. Since it is a data set provided by a public data portal, it is necessary to change the language from Korean to English, in the process of which unnecessary attributes will be excluded from Excel. The preprocessed data can be obtained from the address below. The purpose of this chapter is to find out how many people get on and off during the morning rush hours, so we need to ascertain whether people get on or off; the day of the week; and the number of people getting on and off per hour. As a result, the configurations of the selected data sets are shown below in Table 9-4. You can download the pre-processed passenger data for subway line 1~4 (Seoul Metro) via the address below.

https://drive.google.com/file/d/1fm6UOyZ9rnmVqJLfo9JEJLS0Bt1QG4W9/view?usp=sharing

**Table 9-4**   Preprocessed dataset

| Attribute | Attribute description |
|---|---|
| Type | The type of the getting on or off (On/Off) |
| DoW (Day of Week) | Day of Week (Mon~Sun) |
| Time | The time of getting on and off (05~02) |

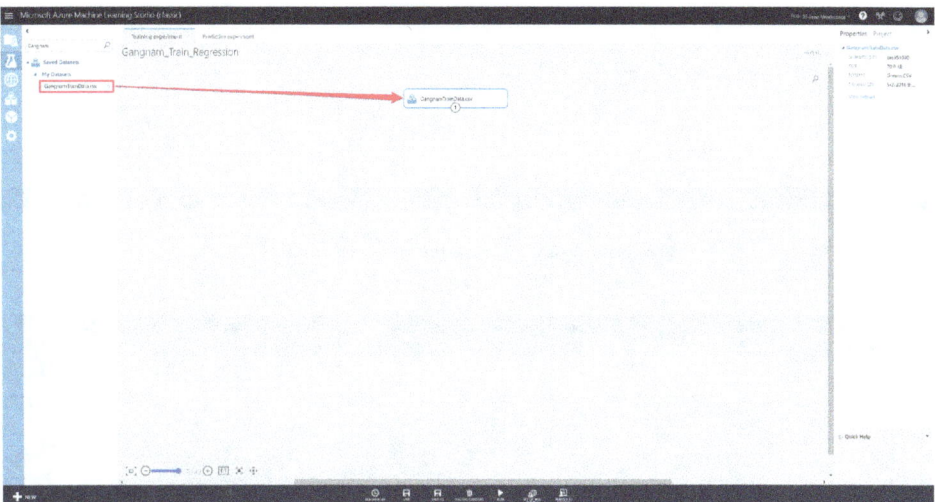

**Figure 9-44**   The data set required for the predictive model

Before building the predictive model, retrieve the preprocessed data and place it as shown in Figure 9-44.

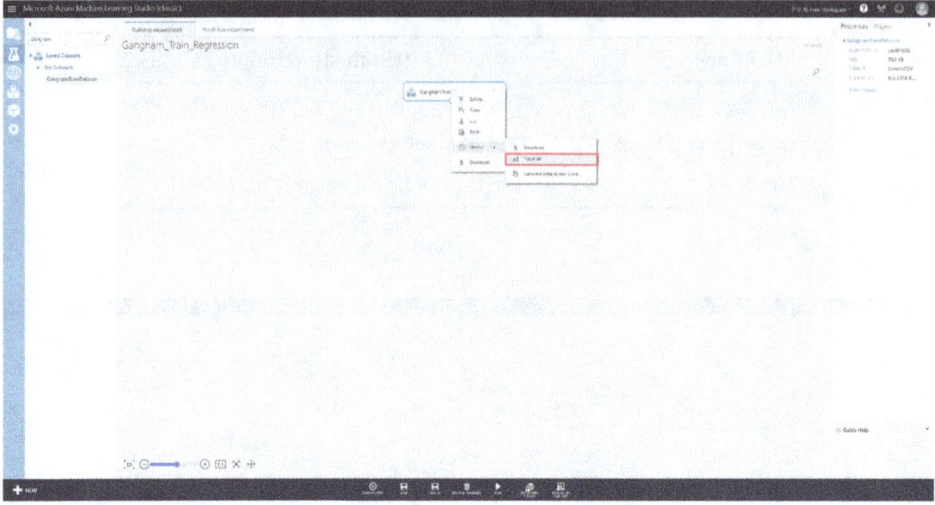

**Figure 9-45**   Check the imported data

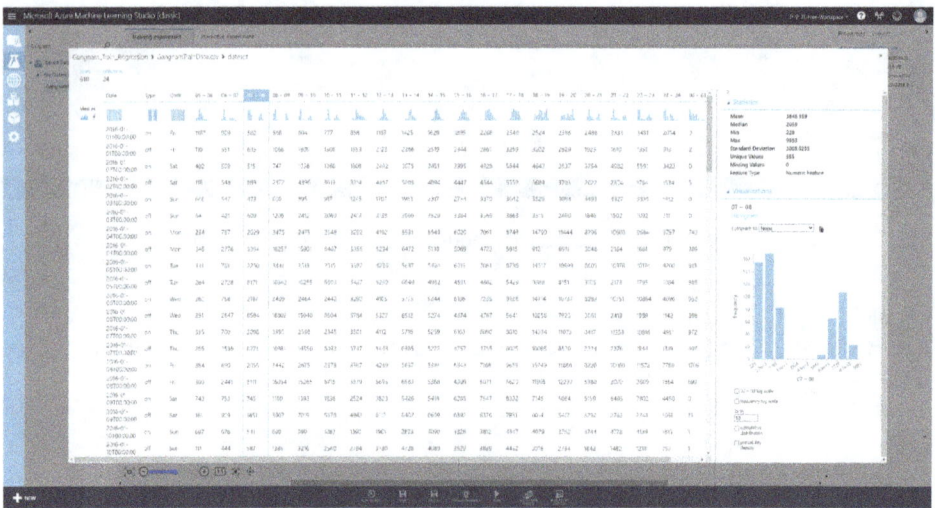

**Figure 9-46**   Check the imported data set distribution

Right-click on the placed data set block to visualize it, as shown in Figure 9-45, and check that the data is intact and that the required data is in place.

Having visualized the data set in Figure 9-46, we can confirm that the data showed up correctly, and that the 'Date' property is not necessary for the production of the prediction model.

Before training a predictive model, you should specify which attributes are for learning.

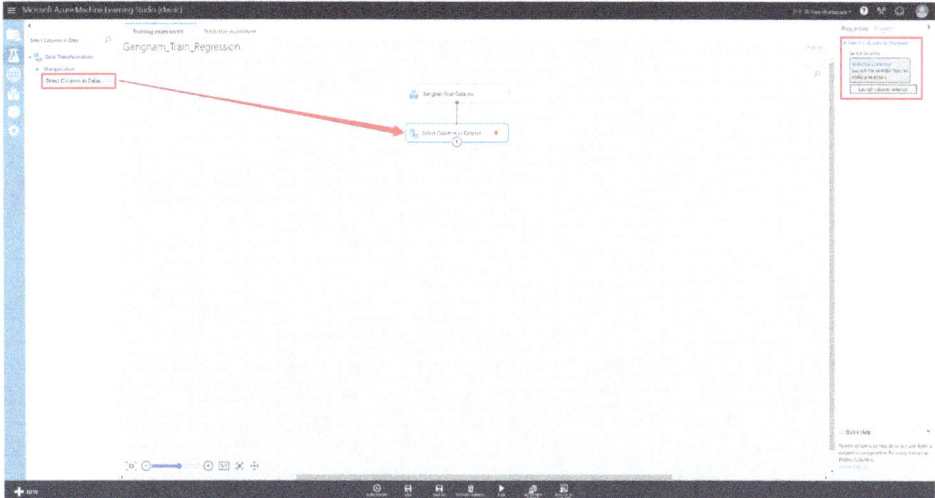

**Figure 9-47**   Attribute selection for training predictive models

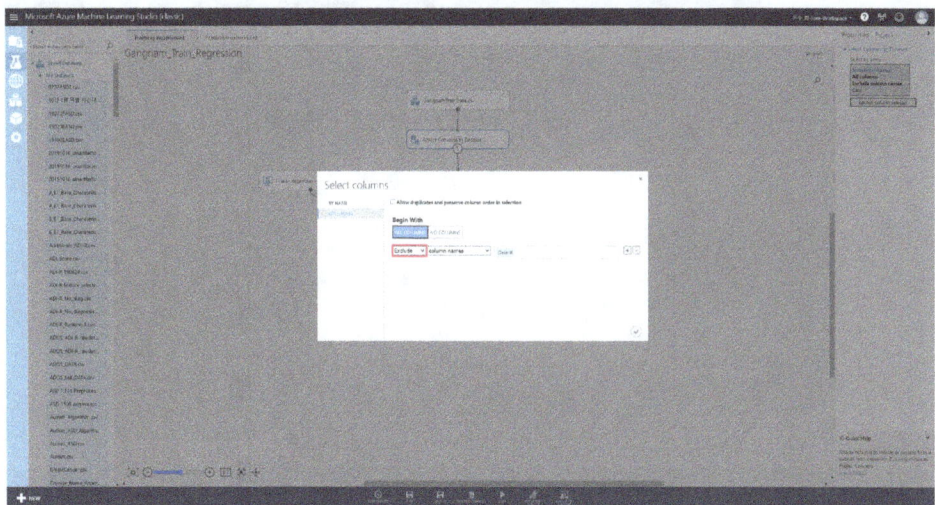

**Figure 9-48**   Exclude unnecessary attributes for training predictive models

Retrieve the "Select Columns in Dataset" block and connect it with the data block, as shown in Figure 9-47 and then click "Launch column selector" on the right to specify which attribute will be used to train the prediction model.

Since we are going to make a predictive model using all the properties except "Date", select "ALL COLUMNS" and exclude "Date" using the exclude function, as shown in Figure 9-48.

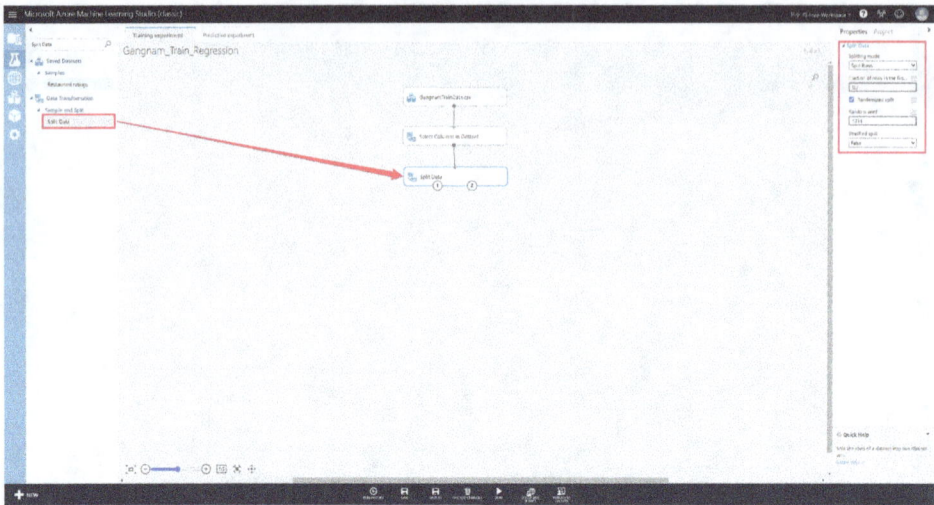

**Figure 9-49**    Data split

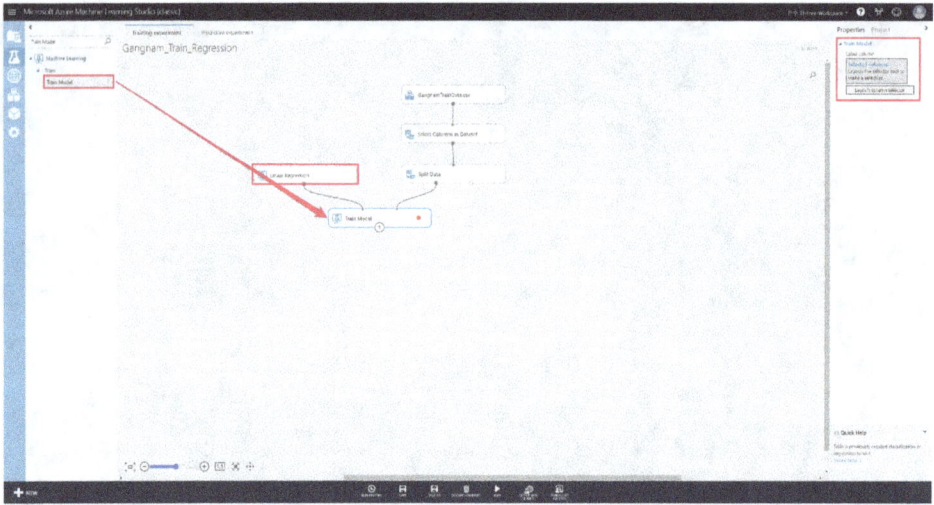

**Figure 9-50**    Placement of "Linear Regression" algorithm block and the "Train Model" block for Predictive Model

If you specified that you are training a predictive model for an attribute, as shown in Figure 9-49, then place a "Split Data" block and segment the data by setting the data for training to 70%, and the data for prediction to 30%.

After segmenting the data for training, place the "Train Model" block for predictive model training and the "Linear Regression" block, which is the algorithm that will be applied to the predictive model, as shown in Figure 9-50.

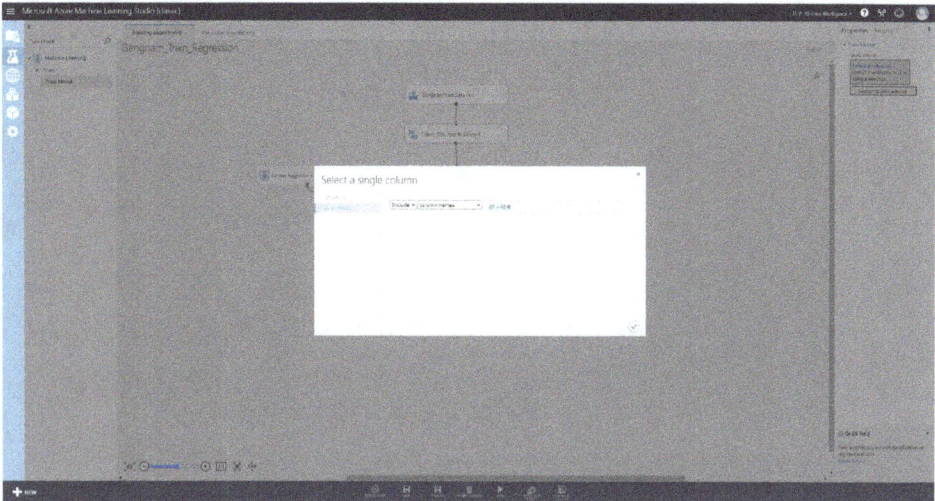

**Figure 9-51**   Select attribute to predict

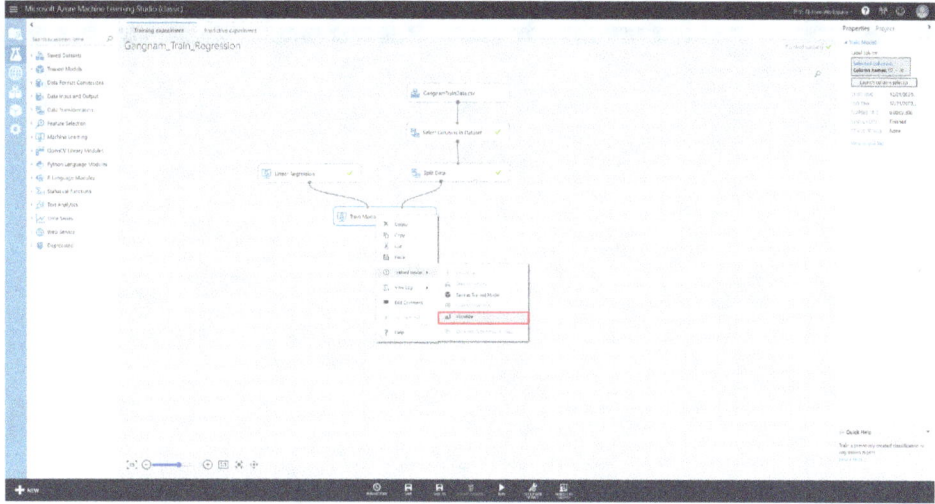

**Figure 9-52**   Check trained prediction model

After placing the blocks, you need to set which properties you will learn by clicking the "Launch column selector" in the "Train Model" block.

The desired value in this chapter is the number of people getting on during the morning rush hours. Therefore, the objective is to predict the number of people between 07:00 and 08:00, so select the 07-08 property, as shown in Figure 9-51.

Click the "Run" button at the bottom, and when the model has completed the learning, check the trained model by visualizing it as shown in Figure 9-52.

**Figure 9-53**    Check the weights by feature in the trained prediction model

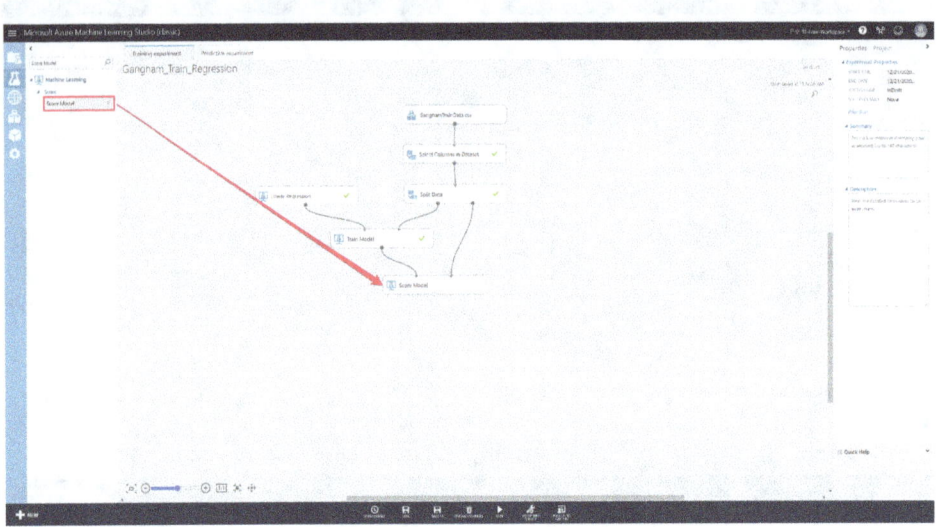

**Figure 9-54**    Placement for the prediction of the trained model

As shown in Figure 9-53, you can check the parameter values of the applied algorithm and the weights learned by each attribute.

Now that the prediction model is complete, place the "Score Model" block for prediction and connect the 30%-split data with the "Train Model" block, as shown in Figure 9-54.

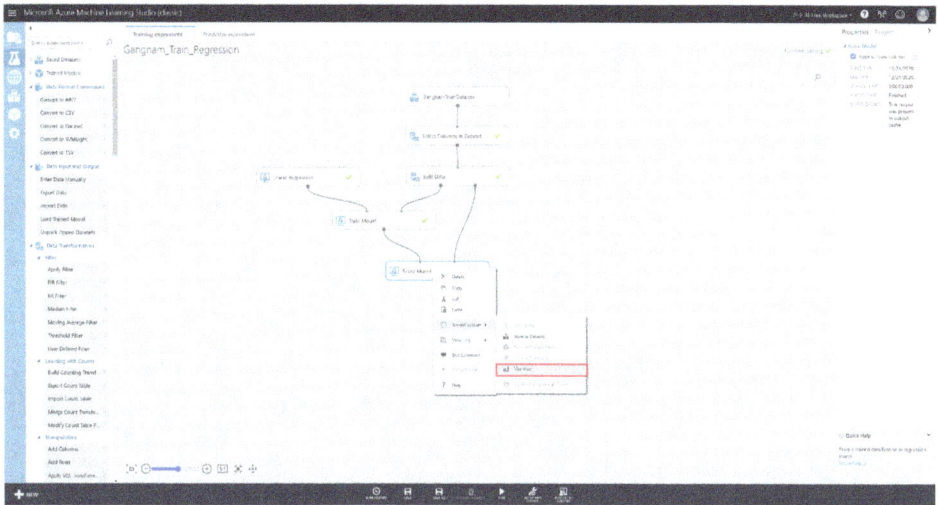

**Figure 9-55**  Check the predicted results

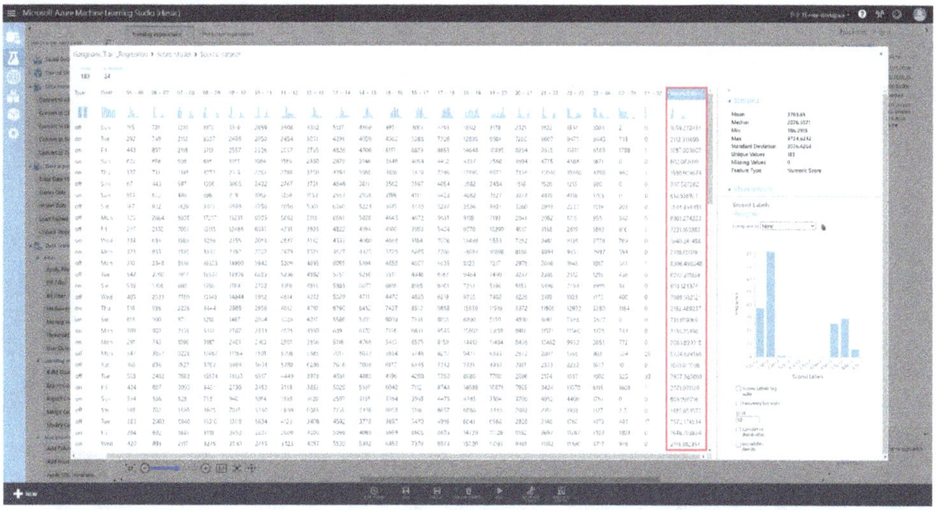

**Figure 9-56**  Check the predicted values

If the prediction has completed after clicking the "Run" button at the bottom, let's check the prediction result through visualization, as shown in Figure 9-55.

As shown in Figure 9-56, you can check the estimated number of people from 07:00 to 08:00.

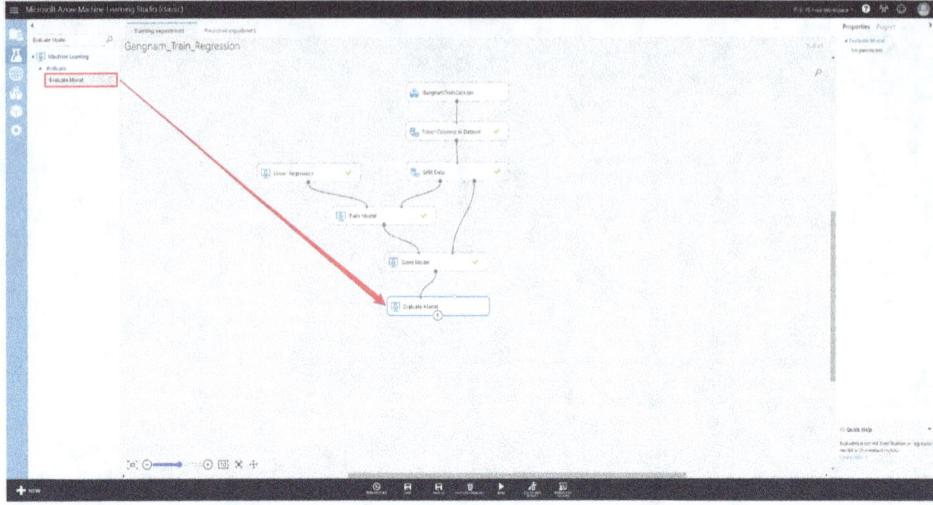

**Figure 9-57**   Placement for verification of predicted results

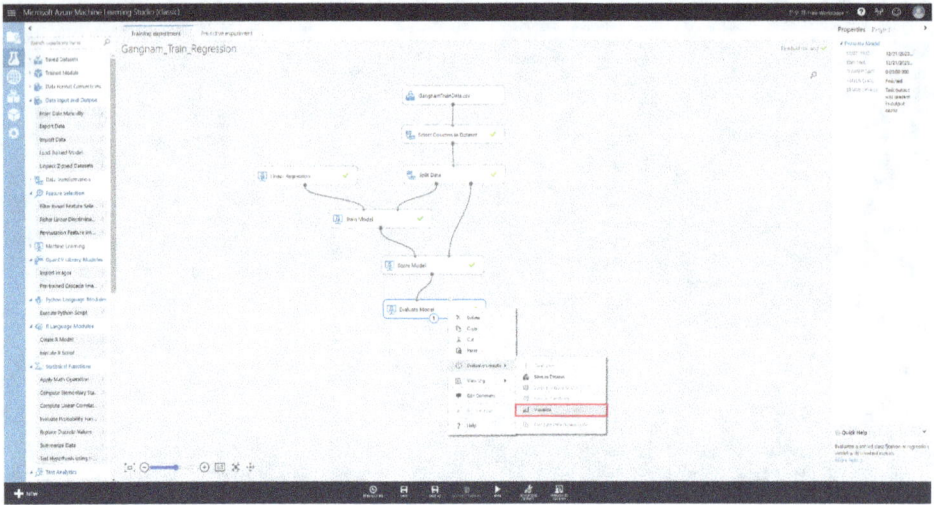

**Figure 9-58**   Confirm the verification results

Now that the prediction is complete, let's place the "Evaluate Model" block as shown in Figure 9-57, and click the "Run" button at the bottom to get the statistical verification value for our value.

When the execution is complete, right-click the "Evaluate Model" block to check it, as shown in Figure 9-58.

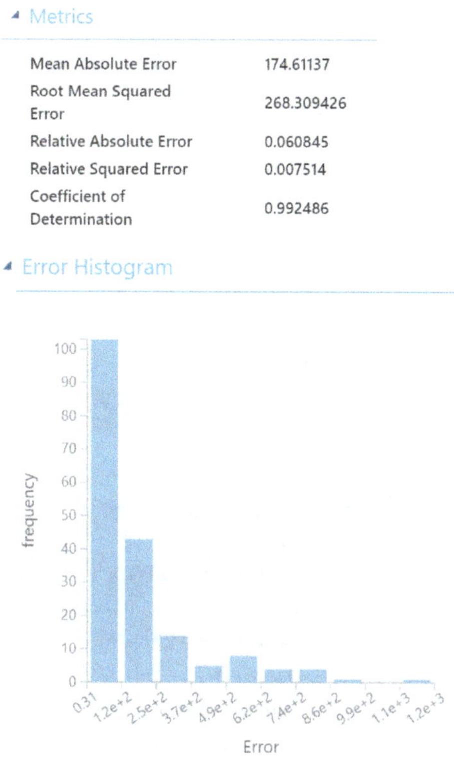

▲ Metrics

| | |
|---|---|
| Mean Absolute Error | 174.61137 |
| Root Mean Squared Error | 268.309426 |
| Relative Absolute Error | 0.060845 |
| Relative Squared Error | 0.007514 |
| Coefficient of Determination | 0.992486 |

▲ Error Histogram

**Figure 9-59**   Evaluation result

Figure 9-59 is the result of evaluating the performance of the predictive model.
Evaluated results for predicted values such as MAE (mean absolute error) and RMSE (root mean square error) can be confirmed.

## 9.5  Heart Disease Prediction

This exercise uses the Two-Class Support Vector Machine and the Multiclass Decision Jungle, as well as the Two Class Decision Jungle, to predict the occurrence of heart disease from the heart disease data set provided by UCI.

The heart disease data can be accessed via the address below.

https://archive.ics.uci.edu/ml/datasets/Heart+Disease

The number of rows in the data set is 303, and the number of attributes is 14, including the label. The attribute name and property description of the data set can be found in Table 9-5 below.

This practice can be found in the AI Gallery. For details and information on importing the project, access the link below.

https://gallery.azure.ai/Experiment/Heart-Disease-Prediction-2

**Table 9-5**   Heart disease dataset

| Attribute name | Attribute explanation |
| --- | --- |
| age | Age in years |
| sex | Sex (1 = male; 0 = female) |
| chestpaintype | Chest pain type<br>– Value 1: typical angina<br>– Value 2: atypical angina<br>– Value 3: non-anginal pain<br>– Value 4: asymptomatic |
| resting_blood_pressure | Resting blood pressure |
| serum_cholesterol | Serum cholesterol in mg/dl |
| fasting_blood_sugar | fbs<br>(fasting blood sugar > 120 mg/dl) (1 = true; 0 = false) |
| resting_ecg | Resting electrocardiographic results<br>– Value 0: normal<br>– Value 1: having ST-T wave abnormality (T wave inversions and/or ST elevation or depression of > 0.05 mV)<br>– Value 2: showing probable or definite left ventricular hypertrophy by Estes' criteria |
| max_heart_rate | Maximum heart rate achieved |
| exercise_induced_angina | Exercise induced angina (1 = yes; 0 = no) |
| st_depression_induced_by_ exercise | ST depression induced by exercise relative to rest |
| slope_of_peak_exercise | Slope: the slope of the peak exercise ST segment<br>– Value 1: upsloping<br>– Value 2: flat<br>– Value 3: downsloping |
| number_of_major_vessel | Ca: number of major vessels (0-3) colored by flouroscopy |
| thal | normal = 3, fixed defect = 6, reversible defect = 7 |
| heart_disease_diag | Diagnosis result<br>(no = 0, yes = 1 ~ 4) |

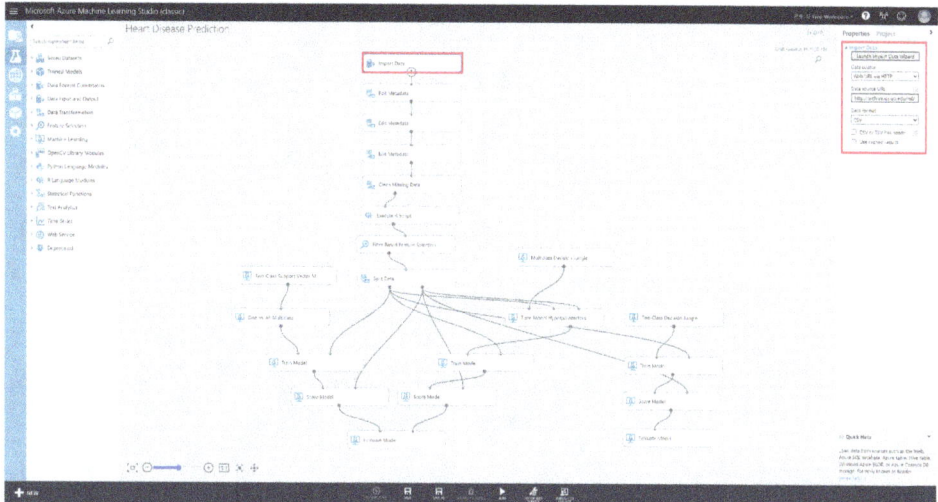

**Figure 9-60** Import dataset

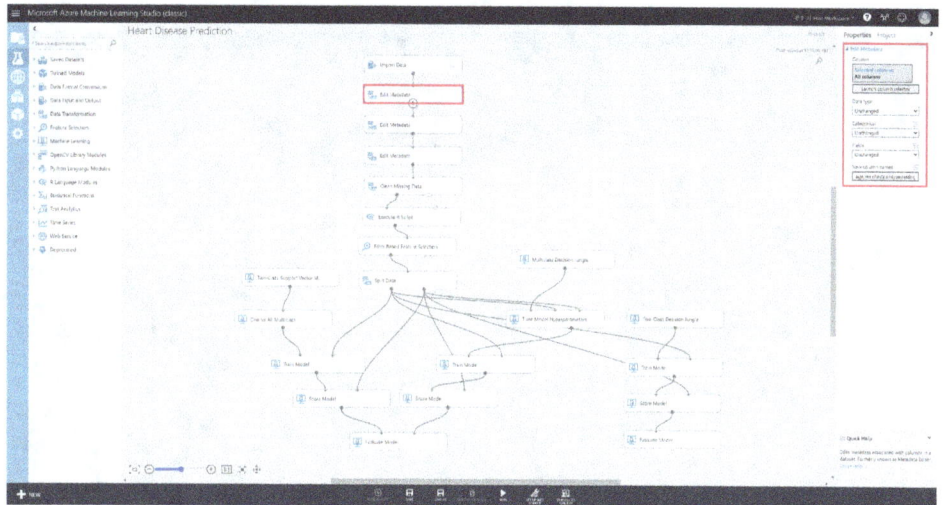

**Figure 9-61** Set Attribute name

Import the UCI dataset via the URL. As shown in Figure 9-60, you can import data by converting it to CSV format using the address below.

http://archive.ics.uci.edu/ml/machine-learning-databases/heart-disease/processed.cleveland.data

The data is retrieved via the URL, but the attribute name is not defined. Write the name of the attribute to define sequentially via the "Edit Metadata" block. Inputted attribute names are entered in order. In Figure 9-61, the attribute names are written sequentially for the column.

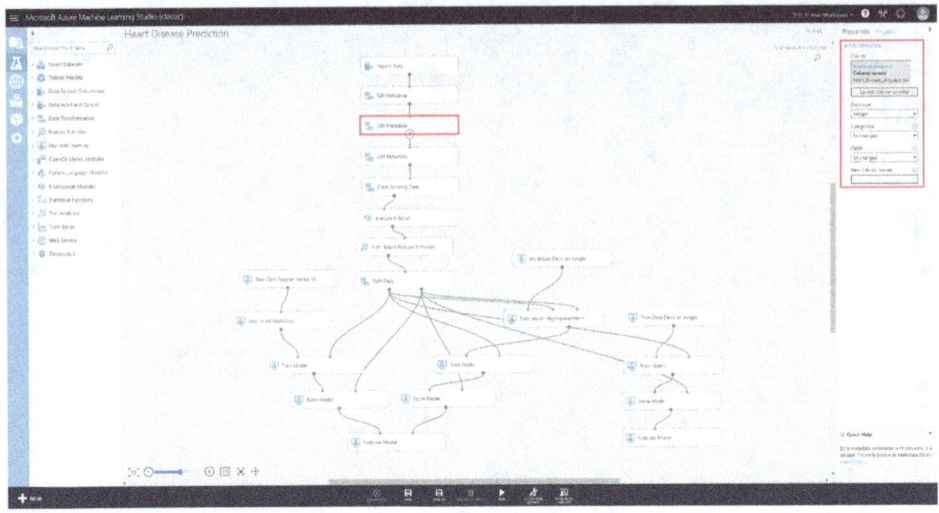

**Figure 9-62**    Data type conversion

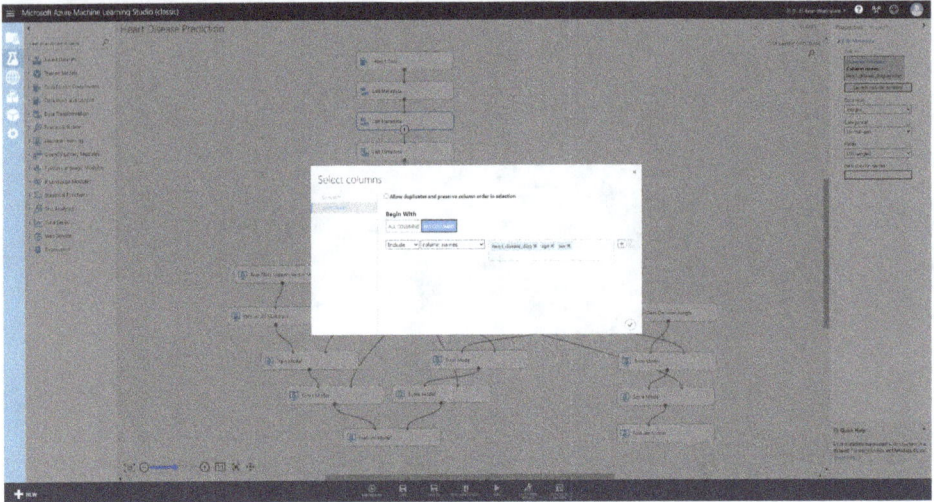

**Figure 9-63**    Select column for conversion

Figure 9-62 shows the conversion of a specified attribute to an integer before training the prediction model.

Figure 9-63 shows how to set the columns to "convert", "heart_disease_diag",""age", and "sex".

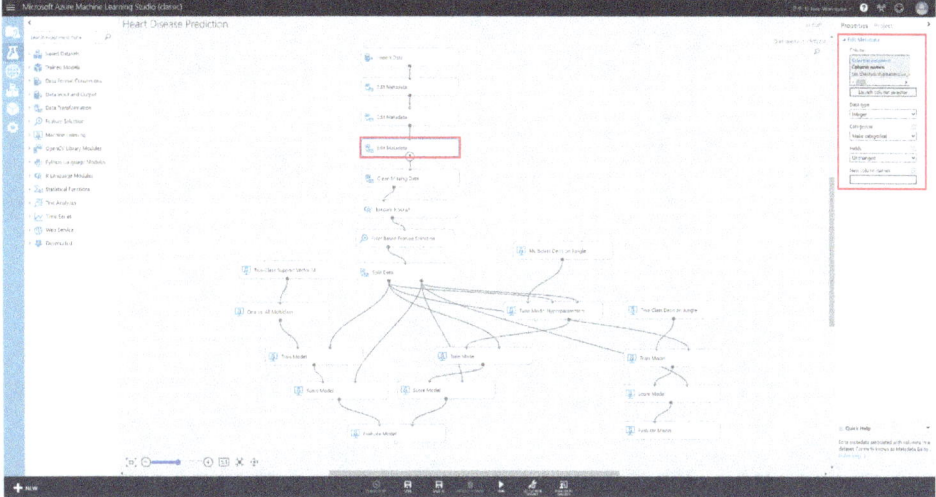

**Figure 9-64** Data type conversion

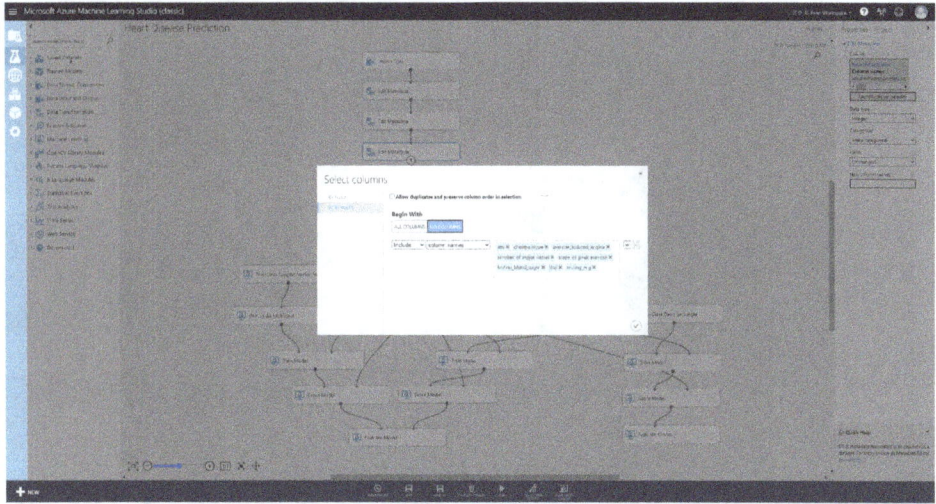

**Figure 9-65** Select column for conversion

Figure 9-64 shows the conversion of a specified attribute to an integer before training the prediction model.

Figure 9-65 shows how to set the columns to convert: "sex", "chestpaintype", "exercise_induces_angina", "number_of_major_vessel", "thal", "slope_of_peak_exercise", "fasting_blood_sugar" and "resting_ecg".

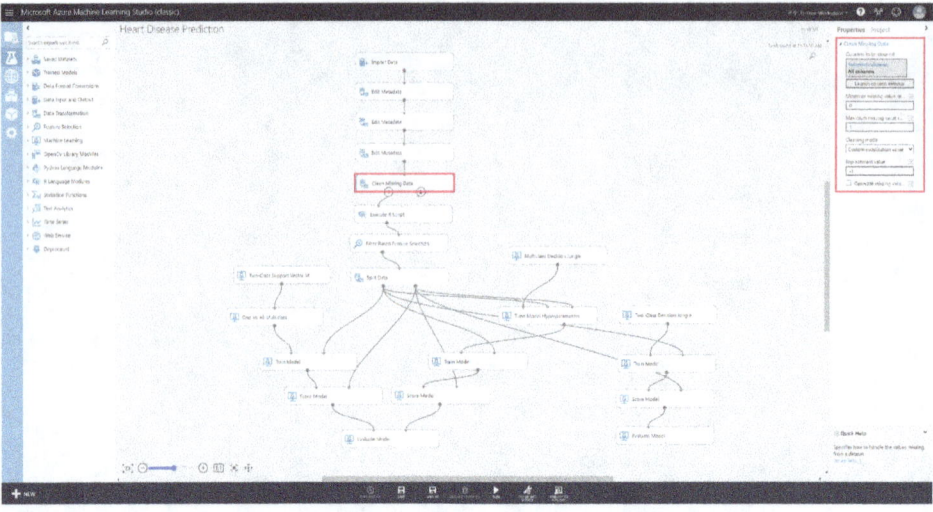

**Figure 9-66**    Preprocessing for missing data

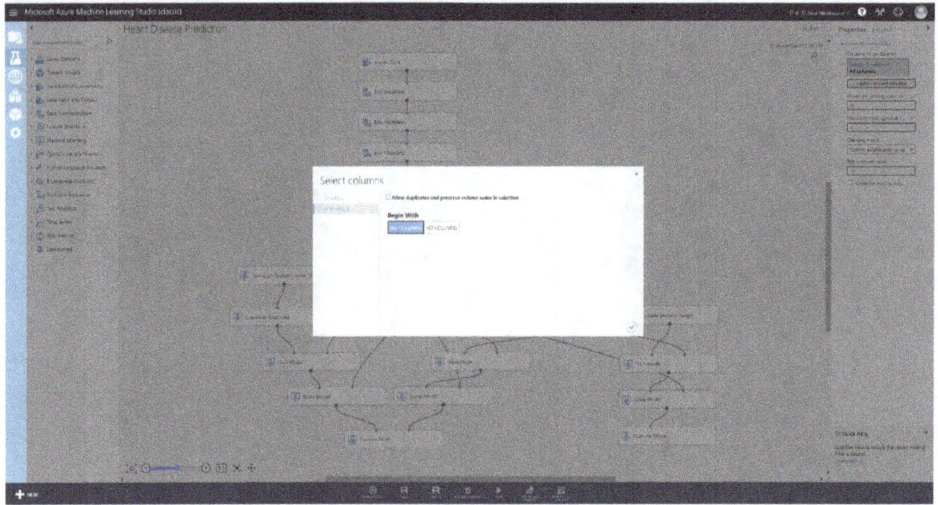

**Figure 9-67**    Preprocessing range for missing data

Preprocessing of the imported data is complete, and all attributes have been searched via the "Clean Missing Data" block to ensure that there is no missing data. Figure 9-66 fills in the missing data with −1.

Figure 9-67 shows the preprocessing of missing data for all attributes.

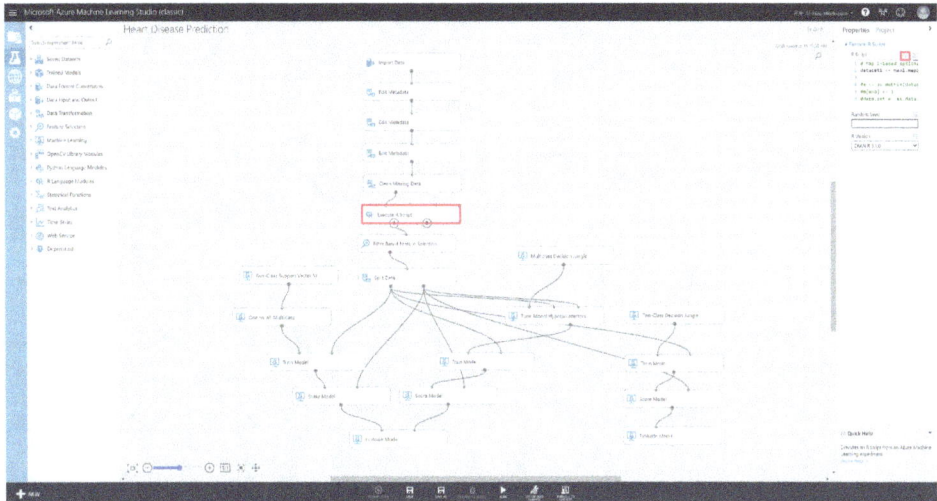

**Figure 9-68** Modifying R Script for label preprocessing

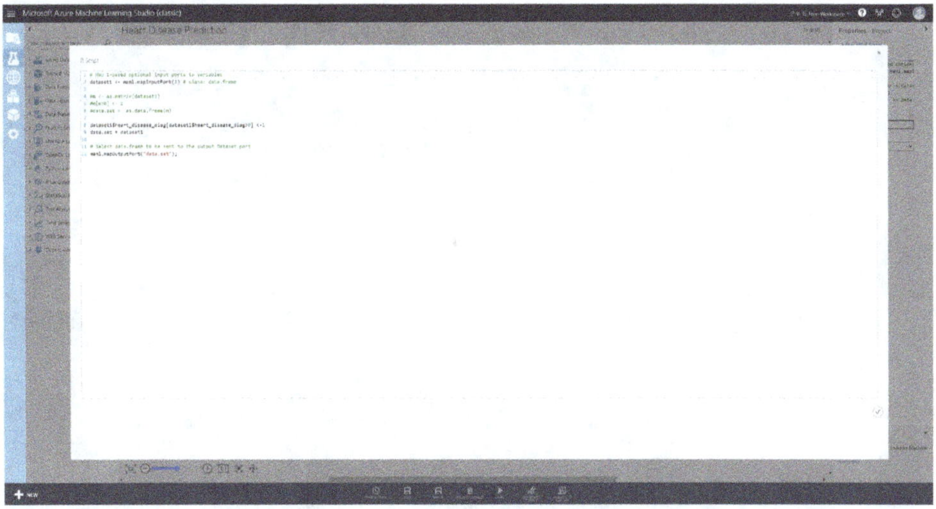

**Figure 9-69** Editing R Script for label preprocessing

When the missing data preprocessing was completed, the R script preprocesses for the "heart_disease_diag" value, which is a label for creating a predictive model, as shown in Figure 9-68.

Figure 9-69 shows an R script that says that if the value of "heart_disease_diag" is bigger than 0, it is replaced with 1. The cardiac diagnosis was changed to 0 (no) and 1 (yes) in order to apply it to the binary algorithm.

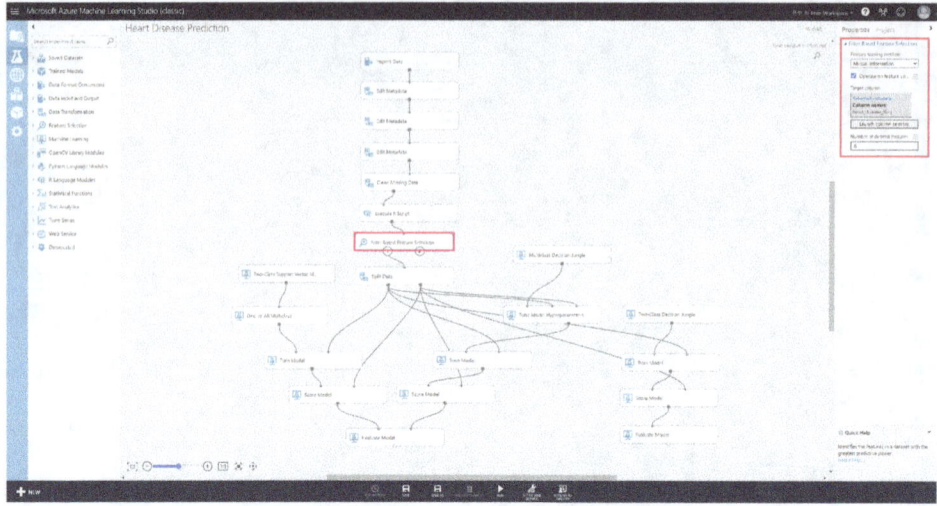

**Figure 9-70**    Set the number of features

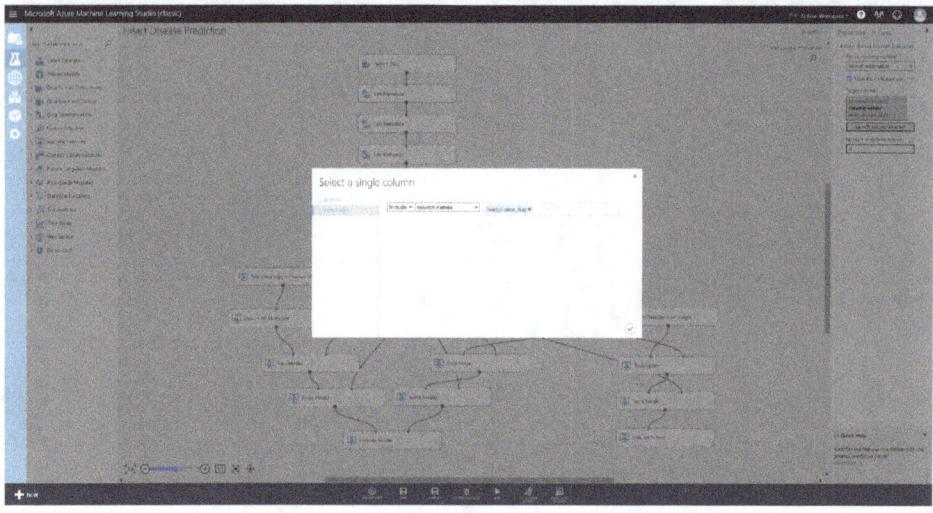

**Figure 9-71**    Set the target column of features

Figure 9-70 shows that the number of attributes to apply to the prediction model training is reduced to eight.

Figure 9-71 is a configuration screen for reducing the eight most influential features to the "heart_disease_diag" label.

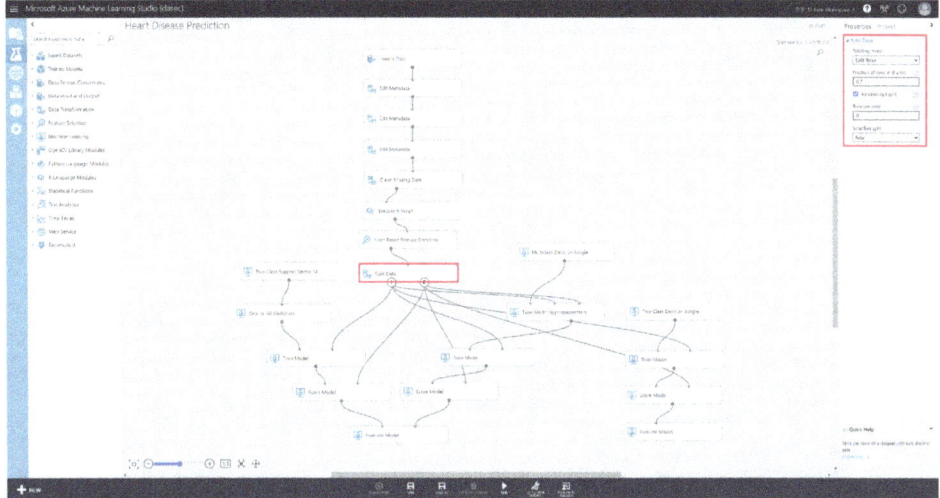

**Figure 9-72** Data split

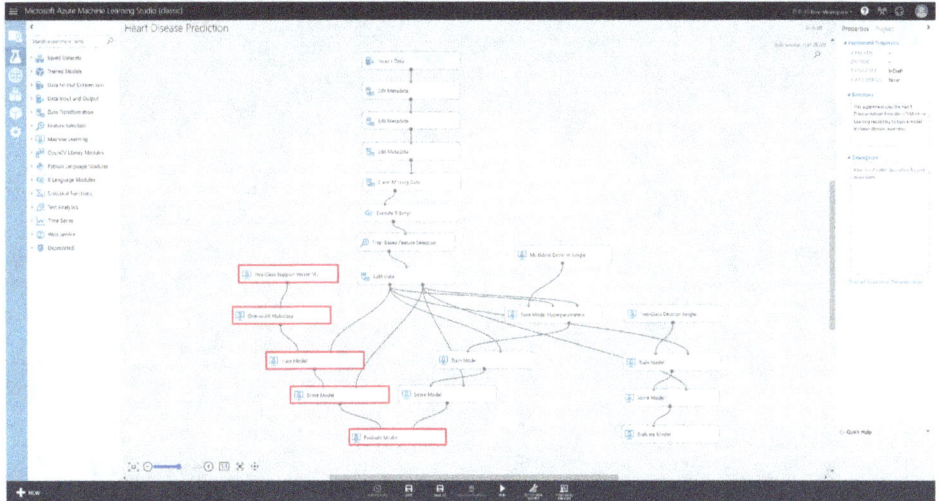

**Figure 9-73** Two-Class SVM prediction model

Figure 9-72 shows the preprocessing of data and the data segmentation for prediction model training and prediction. 70% of the data was allocated for training and 30% of the data was allocated for prediction.

Figure 9-73 shows the arrangement for the predictive model using the Two-Class Support Vector Machine algorithm.

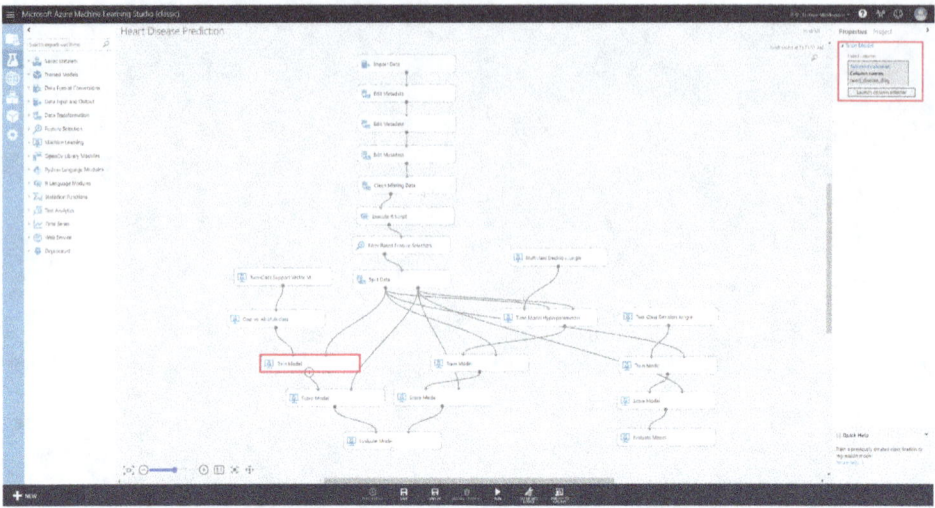

**Figure 9-74**   Set label for prediction model

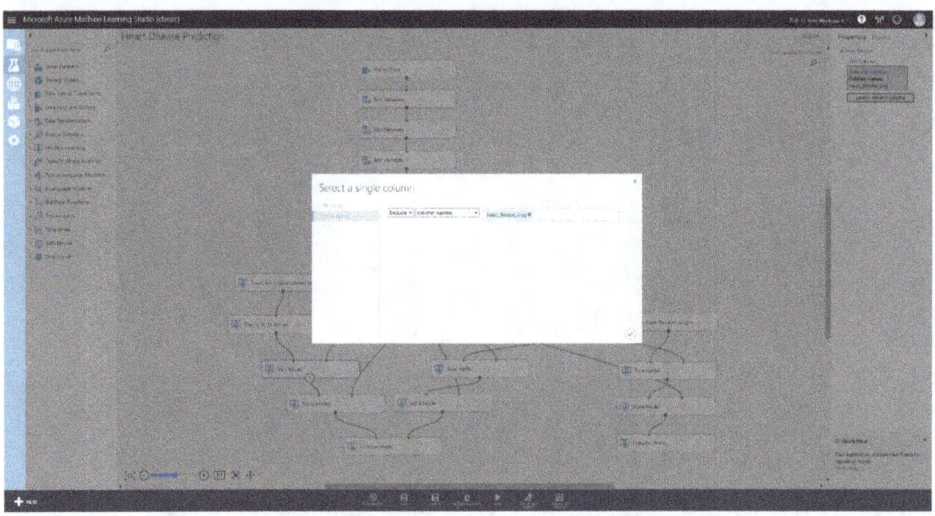

**Figure 9-75**   Set label for prediction model

Figure 9-74 shows how to determine the predicted value of the model by applying the Two-Class Support Vector Machine algorithm.

Figure 9-75 shows the 'heart_disease_diag' column being set as the resulting value of the prediction model via the Launch Column Selector.

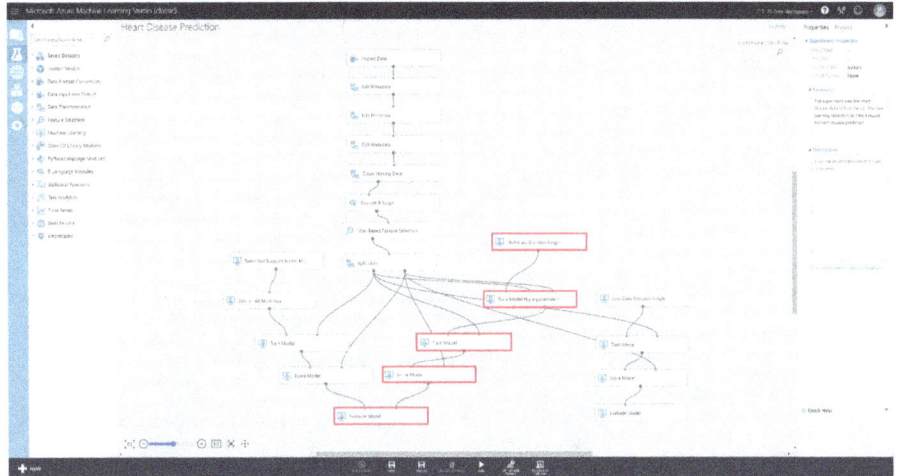

**Figure 9-76** Multiclass Decision Jungle prediction model

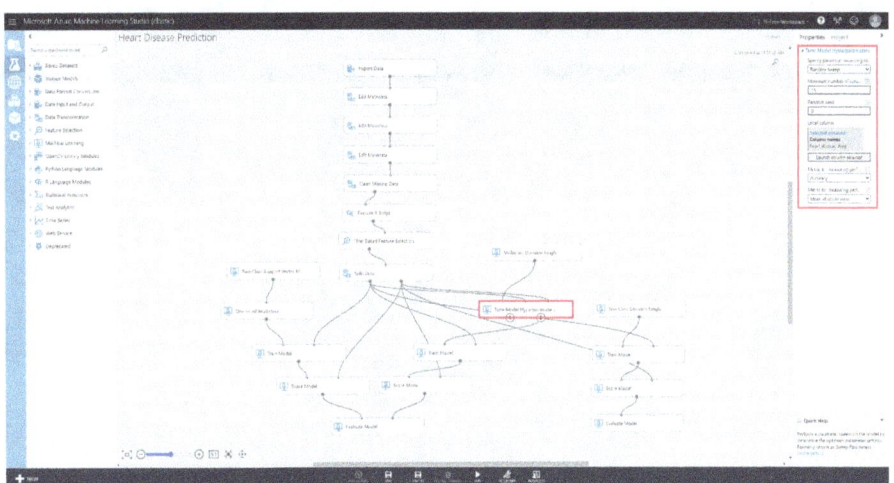

**Figure 9-77** Tune Model Hyperparameters

Figure 9-76 shows the arrangement for the predictive model using the Multiclass Decision Jungle algorithm.

Figure 9-77 shows the empirical selection of "heart_disease_diag", the top parameter for a given algorithm and data set, using the Tune Model Hyperparameter. The accuracy was targeted and applied.

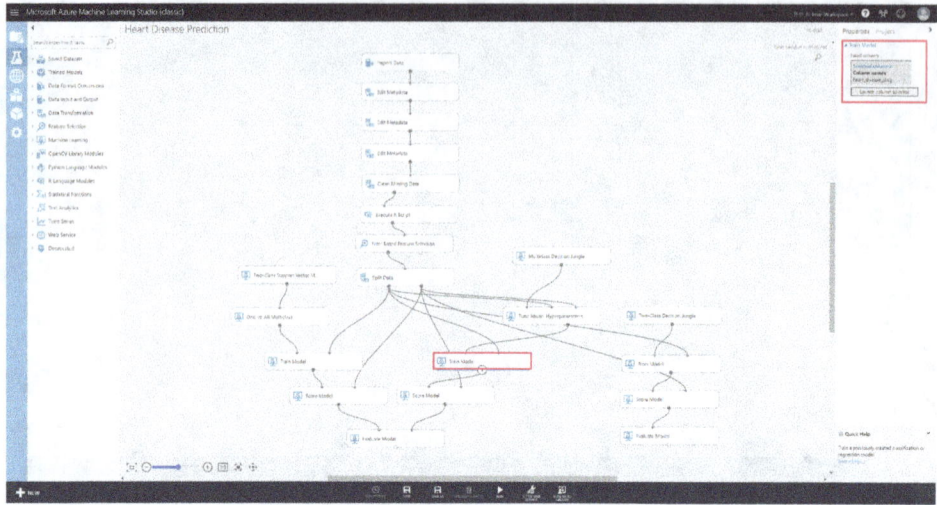

**Figure 9-78**    Set label for prediction model

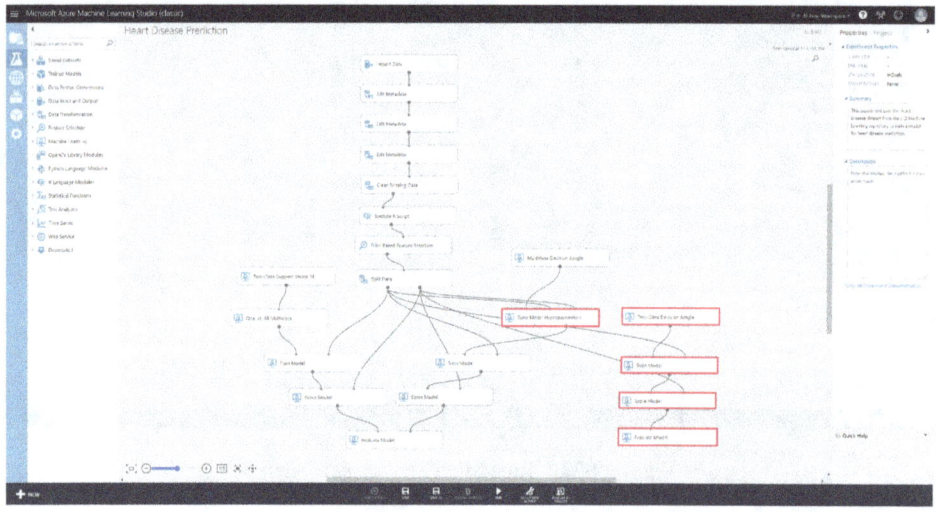

**Figure 9-79**    Two-Class Decision Jungle prediction model

Figure 9-78 shows how to determine the predicted value of the model using the Multiclass Decision Jungle algorithm. Like the Two-Class Support Vector Machine prediction model, it is set as "heart_disease_diag".

Figure 9-79 shows the arrangement for the predictive model using the Two-Class Decision Jungle algorithm.

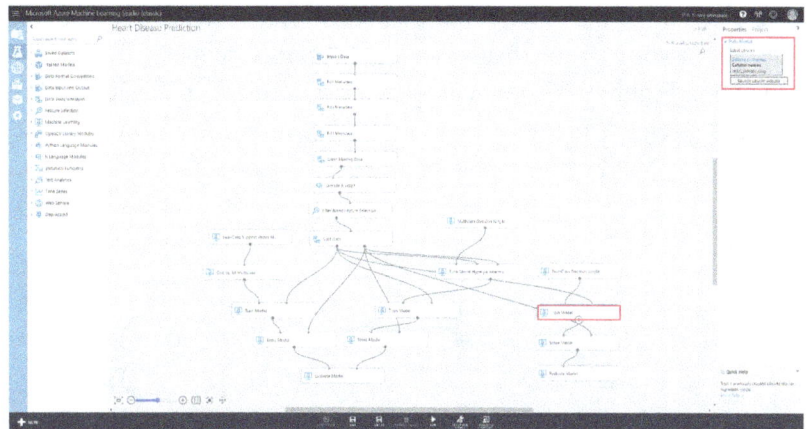

**Figure 9-80**   Set label for prediction model

Heart Disease Prediction ❯ Evaluate Model ❯ Evaluation results

◢ Metrics

| | |
|---|---|
| Overall accuracy | 0.846154 |
| Average accuracy | 0.846154 |
| Micro-averaged precision | 0.846154 |
| Macro-averaged precision | 0.847523 |
| Micro-averaged recall | 0.846154 |
| Macro-averaged recall | 0.834409 |

◢ Metrics

| | |
|---|---|
| Overall accuracy | 0.868132 |
| Average accuracy | 0.868132 |
| Micro-averaged precision | 0.868132 |
| Macro-averaged precision | 0.866919 |
| Micro-averaged recall | 0.868132 |
| Macro-averaged recall | 0.860725 |

◢ Confusion Matrix

Predicted Class

|  | 0 | 1 |
|---|---|---|
| **0** | 90.6% | 9.4% |
| **1** | 23.7% | 76.3% |

Actual Class

◢ Confusion Matrix

Predicted Class

|  | 0 | 1 |
|---|---|---|
| **0** | 90.6% | 9.4% |
| **1** | 18.4% | 81.6% |

Actual Class

**Figure 9-81**   Two-Class Support Machine, Multiclass Decision Jungle evaluation results

Figure 9-80 shows how to determine the predicted value of the model using the Two-Class Decision Jungle algorithm. Like the Two-Class Support Vector Machine prediction model, it is set as "heart_disease_diag".

Figure 9-81 shows a comparison of the prediction results, accuracy, precision, and recall of the Two-Class Support Vector Machine and the Multiclass Decision Jungle. The comparison shows that the accuracy, precision, and recall are higher in the case of the Multiclass Decision Jungle.

Figure 9-82 shows the ROC of the Two-Class Decision Jungle. Looking at the area of the ROC, you can see that it is a well-designed model.

As shown in Figure 9-83, you can check the accuracy, precision and recall of the Two-Class Decision Jungle.

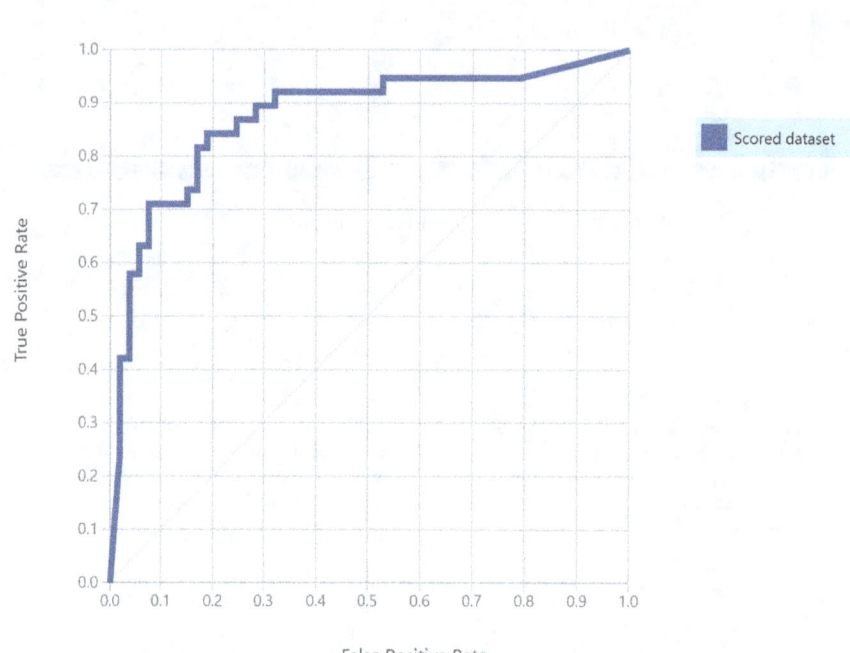

**Figure 9-82**　ROC of Two-Class Decision Jungle

| True Positive | False Negative | Accuracy | Precision | Threshold | AUC |
|---|---|---|---|---|---|
| 27 | 11 | 0.791 | 0.771 | 0.5 | 0.871 |

| False Positive | True Negative | Recall | F1 Score |
|---|---|---|---|
| 8 | 45 | 0.711 | 0.740 |

| Positive Label | Negative Label |
|---|---|
| 1 | 0 |

| Score Bin | Positive Examples | Negative Examples | Fraction Above Threshold | Accuracy | F1 Score | Precision | Recall | Negative Precision | Negative Recall | Cumulative AUC |
|---|---|---|---|---|---|---|---|---|---|---|
| (0.900,1.000] | 13 | 1 | 0.154 | 0.714 | 0.500 | 0.929 | 0.342 | 0.675 | 0.981 | 0.000 |
| (0.800,0.900] | 5 | 1 | 0.220 | 0.758 | 0.621 | 0.900 | 0.474 | 0.718 | 0.962 | 0.008 |
| (0.700,0.800] | 4 | 0 | 0.264 | 0.802 | 0.710 | 0.917 | 0.579 | 0.761 | 0.962 | 0.008 |
| (0.600,0.700] | 2 | 1 | 0.297 | 0.813 | 0.738 | 0.889 | 0.632 | 0.781 | 0.943 | 0.019 |
| (0.500,0.600] | 3 | 5 | 0.385 | 0.791 | 0.740 | 0.771 | 0.711 | 0.804 | 0.849 | 0.084 |
| (0.400,0.500] | 1 | 1 | 0.407 | 0.791 | 0.747 | 0.757 | 0.737 | 0.815 | 0.830 | 0.098 |
| (0.300,0.400] | 4 | 3 | 0.484 | 0.802 | 0.780 | 0.727 | 0.842 | 0.872 | 0.774 | 0.145 |
| (0.200,0.300] | 3 | 10 | 0.626 | 0.725 | 0.737 | 0.614 | 0.921 | 0.912 | 0.585 | 0.315 |
| (0.100,0.200] | 1 | 9 | 0.736 | 0.637 | 0.686 | 0.537 | 0.947 | 0.917 | 0.415 | 0.473 |
| (0.000,0.100] | 2 | 22 | 1.000 | 0.418 | 0.589 | 0.418 | 1.000 | 1.000 | 0.000 | 0.871 |

**Figure 9-83**　Evaluation result of Two-Class Decision Jungle

## 9.6 Find Similar Companies Using K-Means Clustering

This practice session uses the Wikipedia articles dataset to apply a K-Means algorithm in order to predict similarities between companies.

This practice is taken from Microsoft's AI Gallery. For more information on importing the project, access the links below.

https://gallery.azure.ai/Experiment/Clustering-Find-similar-companies-6

The number of data in the Wikipedia articles dataset is less than 500. The Wikipedia articles dataset of the project provided is preprocessed. The preprocessing included:

– Removing wiki formatting
– Removing non-alphanumeric characters
– Converting all text to lowercase
– Adding company categories, where known

The Wikipedia articles dataset consists of three attributes: Title, Category, and Text, as shown in Figure 9-84.

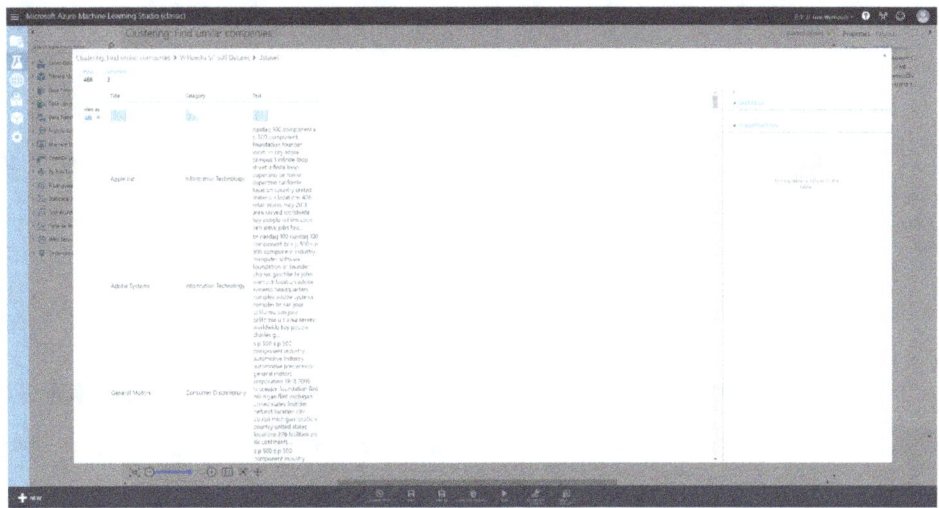

**Figure 9-84**   Wikipedia articles dataset

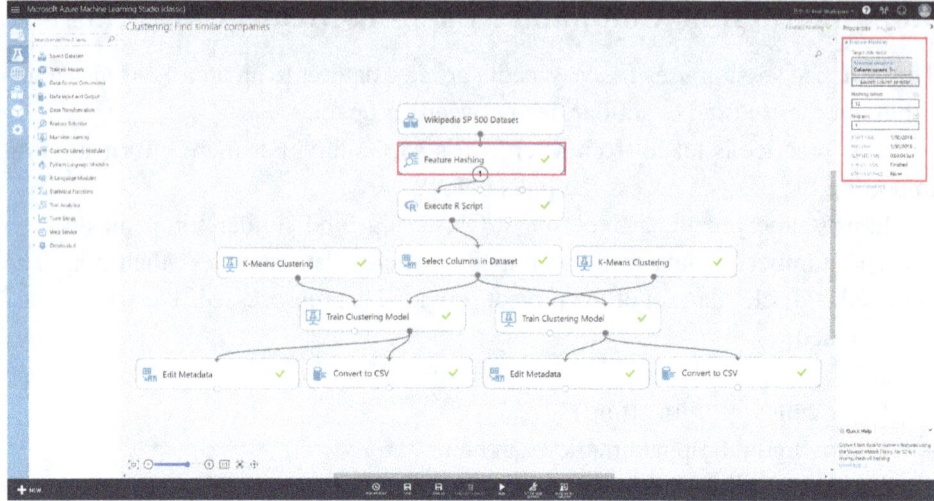

**Figure 9-85**    Feature hashing

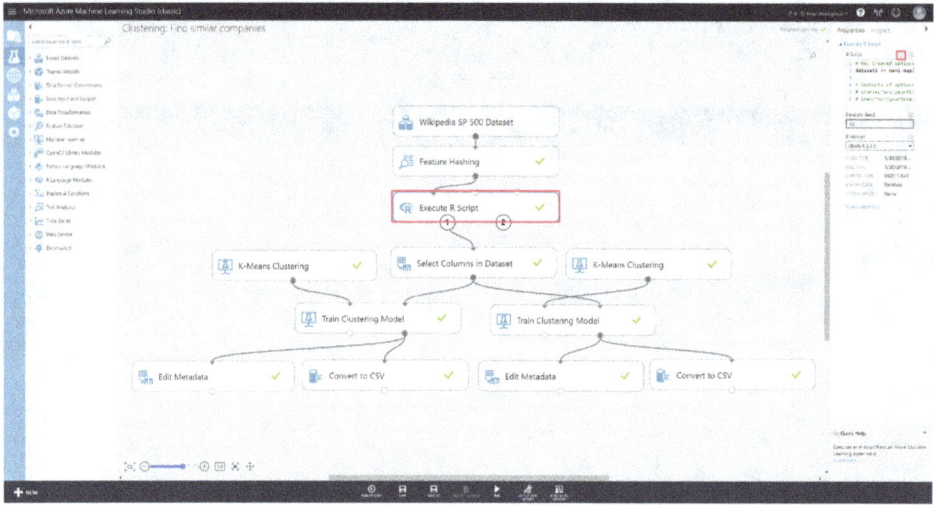

**Figure 9-86**    R Script for implementation PCA

Feature hashing is a method used in the text analysis model that converts text to integers. In Figure 9-85, we can see that the attribute text is selected as the target.

Figure 9-86 shows the R Script for implementing Principal Component Analysis (PCA). Let's look at the source by clicking the "Script modify" button in the upper right corner.

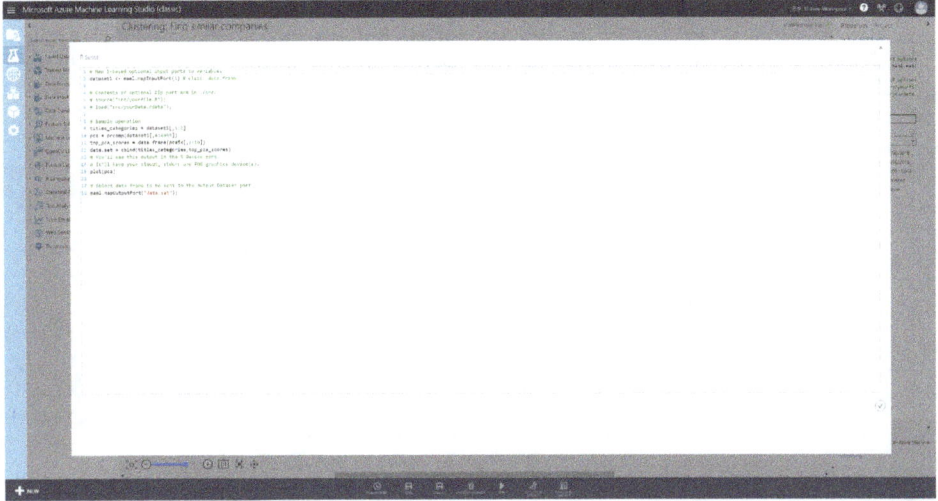

**Figure 9-87** R Script source for implementation PCA

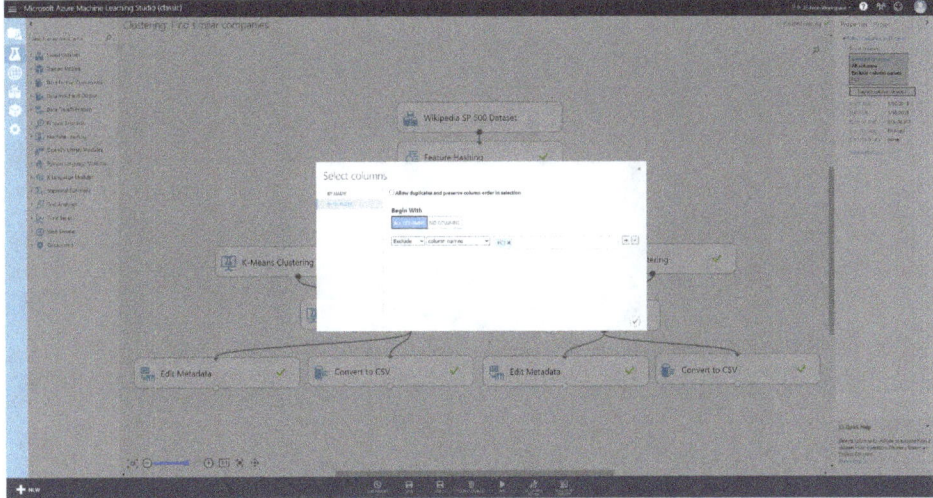

**Figure 9-88** Data attribute selection for training

You can see the R Script source, as shown in Figure 9-87. The number of dimensions of the value of the existing text attribute is high, and the PCA has been applied to reduce it to 10 dimensions.

Select the attribute to be used for learning by excluding PC1, as shown in Figure 9-88.

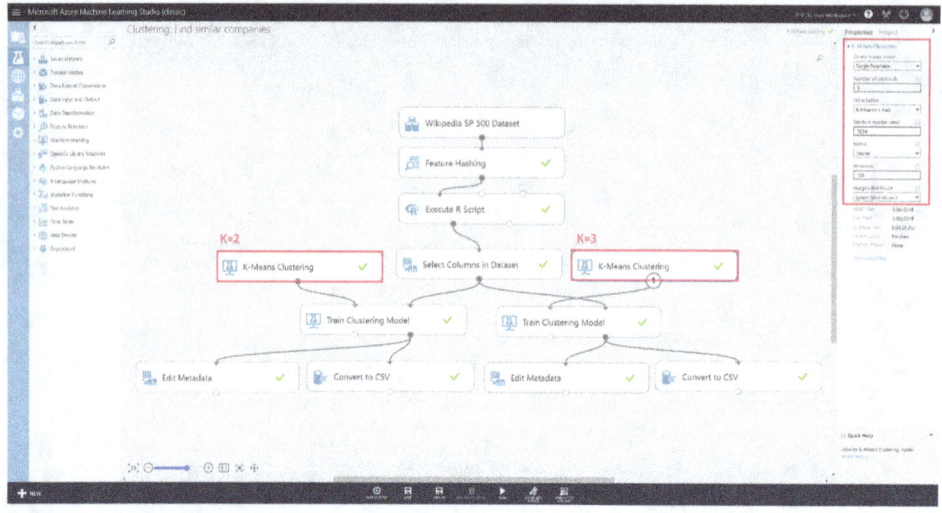

**Figure 9-89**   Set the number of clusters in the K-Means algorithm.

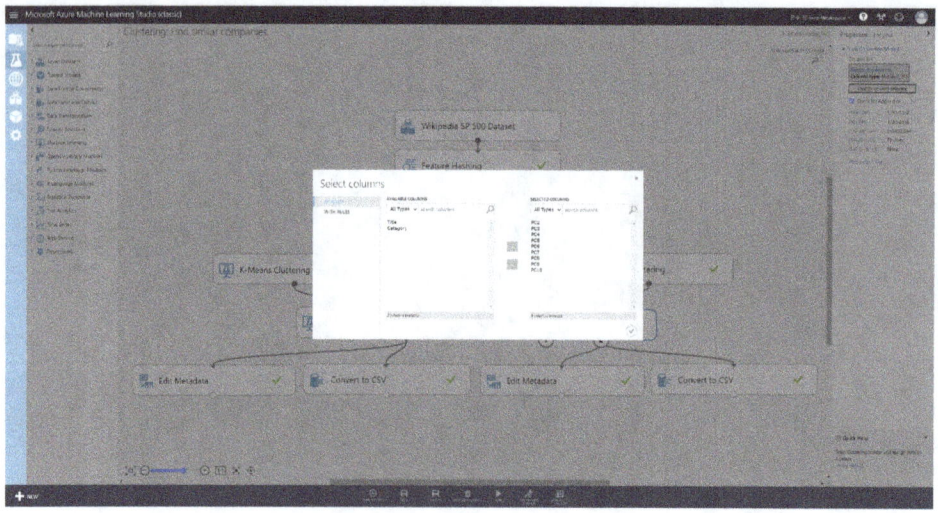

**Figure 9-90**   Train Clustering Model

There are two K-Means algorithms for the model in Figure 9-89. The number of clusters in the left K-Means algorithm is two, and the number of clusters in the right K-Means algorithm is three.

Unlike what we learned earlier, the cluster model uses a different block, the "Train Clustering Model" block. Figure 9-90 shows that clustering is applied via attributes PC2 ~ PC10.

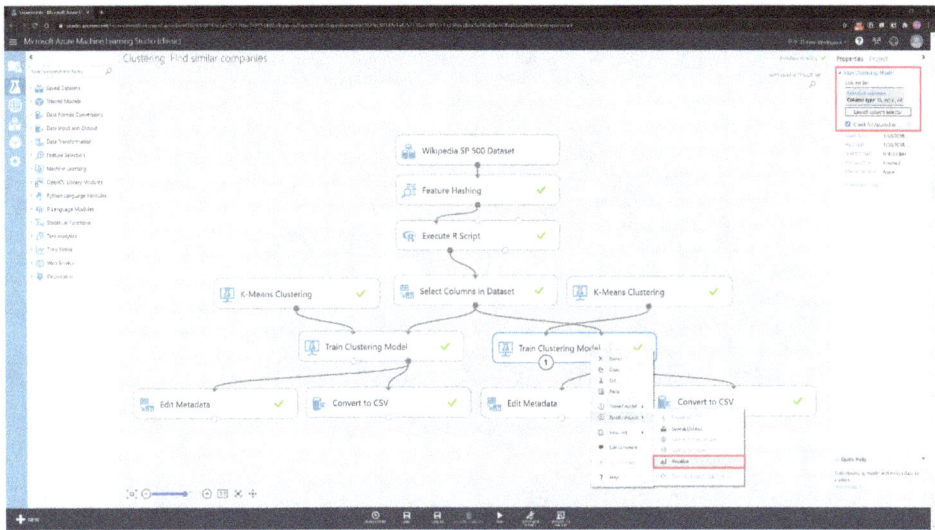

**Figure 9-91**  Clustering result visualization

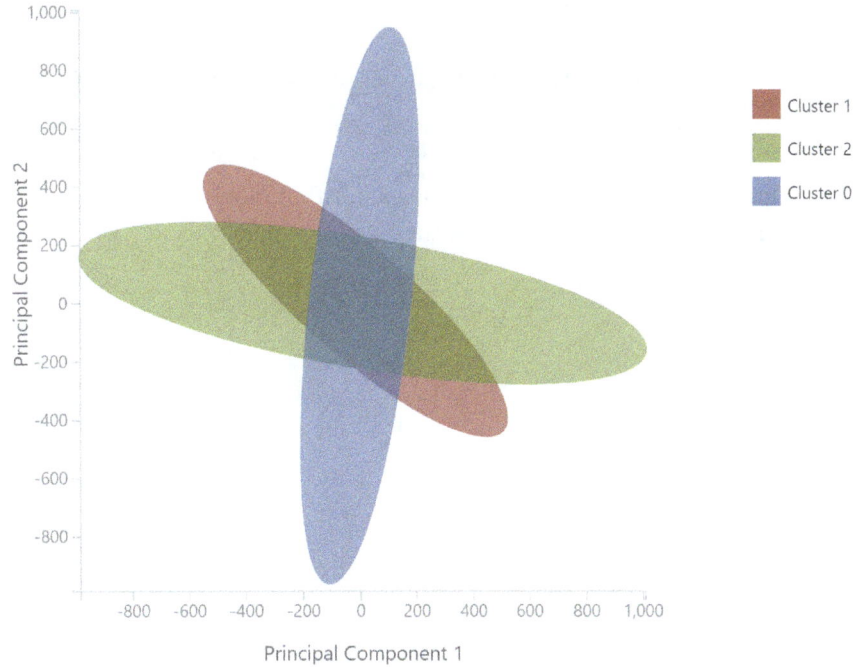

**Figure 9-92**  K-Means result visualization (right)

Click "Run" at the bottom to complete the lesson. Then right-click on the "Train Clustering Model" block, as shown in Figure 9-91, to see the visualization.

The number of clusters in the K-Means algorithm block on the right is defined as three. Figure 9-92 shows three clusters successfully appearing.

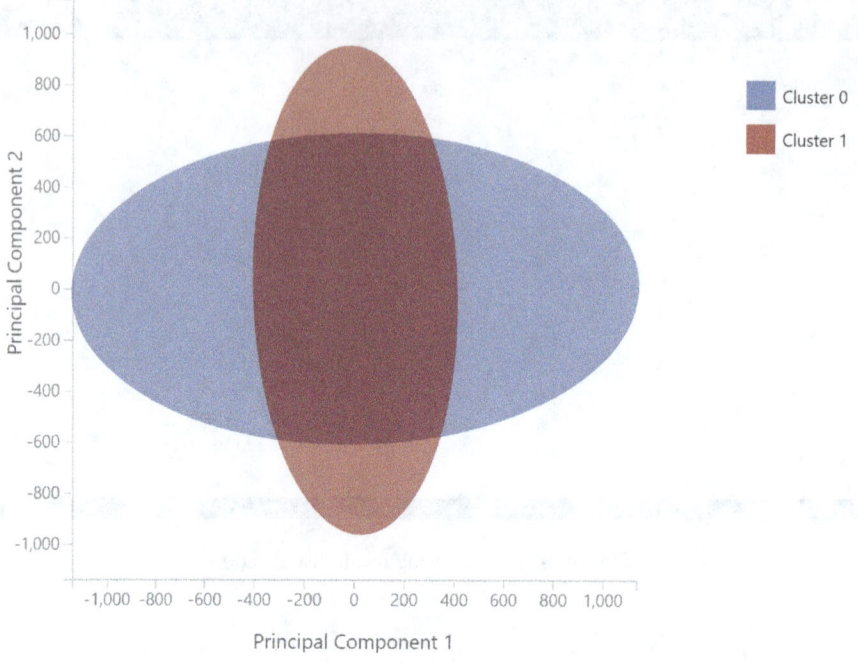

**Figure 9-93**   K-Means result visualization (left)

The number of clusters in the K-Means algorithm block on the left is defined as two. Figure 9-93 shows two clusters successfully appearing.

## 9.7 Practice Questions

Q1. Describe how to preprocess missing data.

Q2. Explain the reason for splitting the data in the prediction model creation process.

# Part III

# Visualization

# Chapter 10

# Visualization

## 10.1 Definition of Visualization

In Wikipedia, Visualization is defined as "any technique for creating images, diagrams, or animations to communicate a message". The most important point in data visualization is the communication of a message. This message is the information gleaned through data visualization. Especially in this age, data visualization is attracting a great deal of attention. There are many reasons for this, but the fact that the visual function is the most powerful information receptor among the human sensory organs, and that the response information receives depends on how it is transmitted are the most significant.

## 10.2 Purpose and Function of Data Visualization

### 10.2.1 *Purpose of Data Visualization*

(1) **Data analysis**
While data visualization is often used to convey messages, it can be used functionally. For example, you can visualize in a certain way to easily check if there is outlier data. In addition, you may draw a Q-Q plot to see if the numerical data follow a normal distribution.

(2) **Share data analysis results**
Visualization methods should be used to share data analysis results as much as possible to easily convey the meaning of the data to others, or to communicate the processed information from the data.

In general, points, lines, legends, titles, and units can be added. In this way, additional information can be expressed so that errors are avoided when plots are interpreted. This also helps the viewer to identify information more quickly.

## 10.2.2 *Function of Data Visualization*

### (1) **Communication**

Communication is a function that delivers information in data quickly and easily, which is possible through visualization. You can maximize the function of communication by making a plot with the user in mind.

### (2) **Discovery**

Discovery is about presenting information to identify new facts. In other words, it is not information that one already knows, but information that delivers new facts from unknown data. In particular, it is a function frequently used in research fields.

### (3) **Insight**

Insight is the expression of information in order to gain a better understanding in cases where information may be hidden, or already known.

## 10.3 Practice Questions

Q1. Explain the definition of visualization.

Q2. Explain the purpose of visualization.

Q3. Explain the function of visualization.

# Chapter 11

# Visualization with Power BI

## 11.1 Introduction of Power BI

In addition to successfully building predictive models, it is also important to visualize the results well. Even a model that shows good performance predictions may not convey meaning unless the results are easily explained. Data visualization helps data scientists easily explain the results of predictive models to stakeholders or end users. Power BI, a business analytics solution, lets you visualize data, share insights with your organization, or import data into apps or websites.

## 11.2 Download and log in to Power BI Desktop

To log in to Power BI, users need a work or school email address. Personal e-mail accounts such as Gmail or Hotmail cannot be used. Go to https://powerbi.micro-soft.com/en-us/ to download Power BI Desktop.

Click "START FREE", as shown in Figure 11-1 to proceed with the download.

Click the "DOWNLOAD FREE" button in Figure 11-2 to run the given MSI installation file. If you want to install a different language, or a 32-bit or 64-bit version of the program, click on "ADVANCED DOWNLOAD OPTIONS" to get the installation file.

Fill out the form shown in Figure 11-3 to create an ID and log in. If you already have a Microsoft account, click the "Sign in" button below to log in.

Log in by entering your Microsoft account in the pop-up, as shown in Figure 11-4.

Figure 11-5 shows a successful log in.

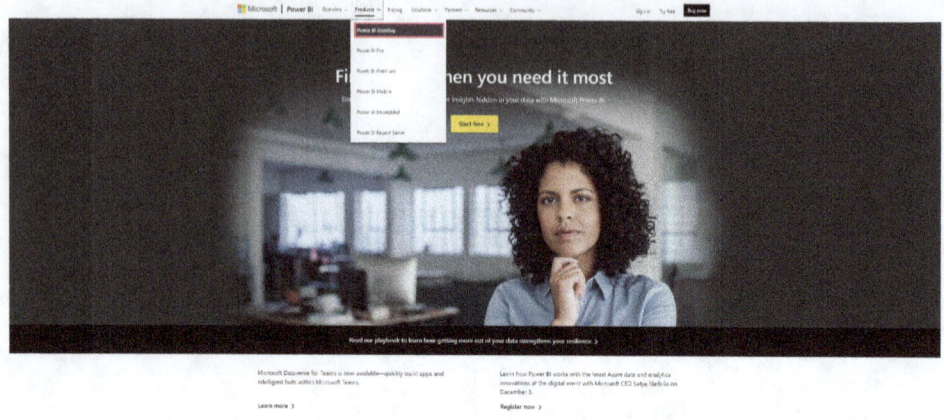

Create a data-driven culture with business intelligence for all

**Figure 11-1**   Power BI Official page

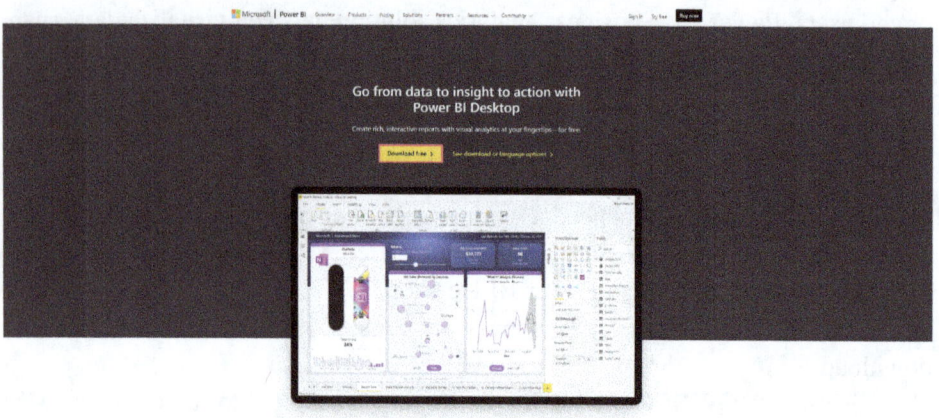

**Figure 11-2**   Download page

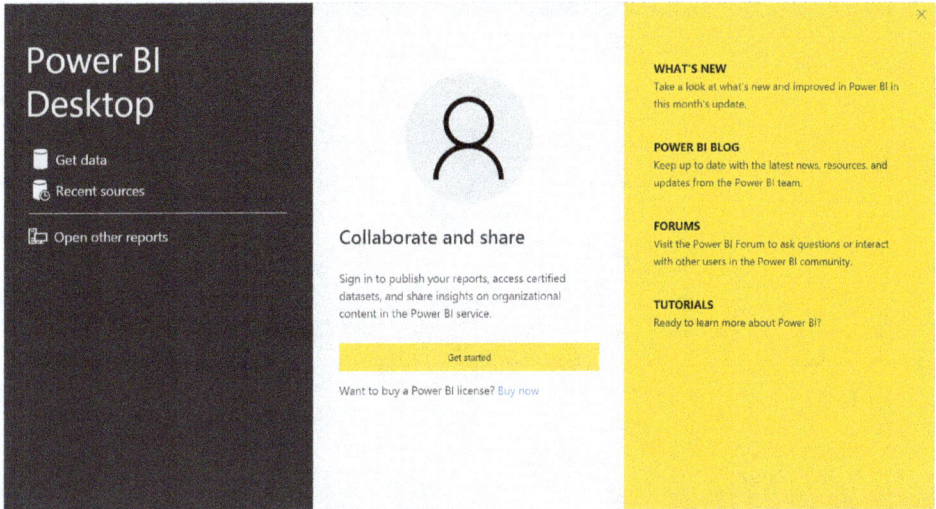

**Figure 11-3** Sign in pop-up

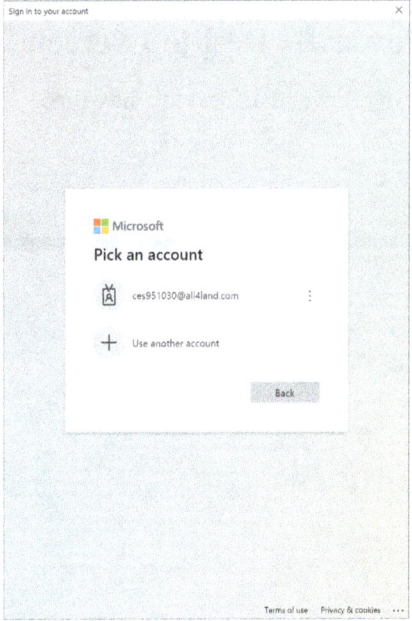

**Figure 11-4** Sign in Pop-up

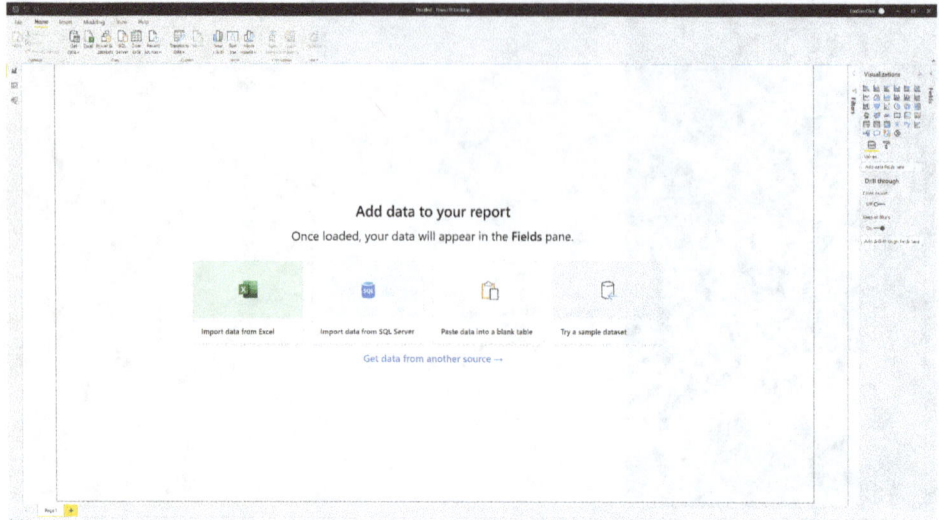

**Figure 11-5**    Log in complete

## 11.3  Configure Power BI Desktop Screen

As shown in Figure 11-6, Power BI Desktop has three views: Report, Data and Model.

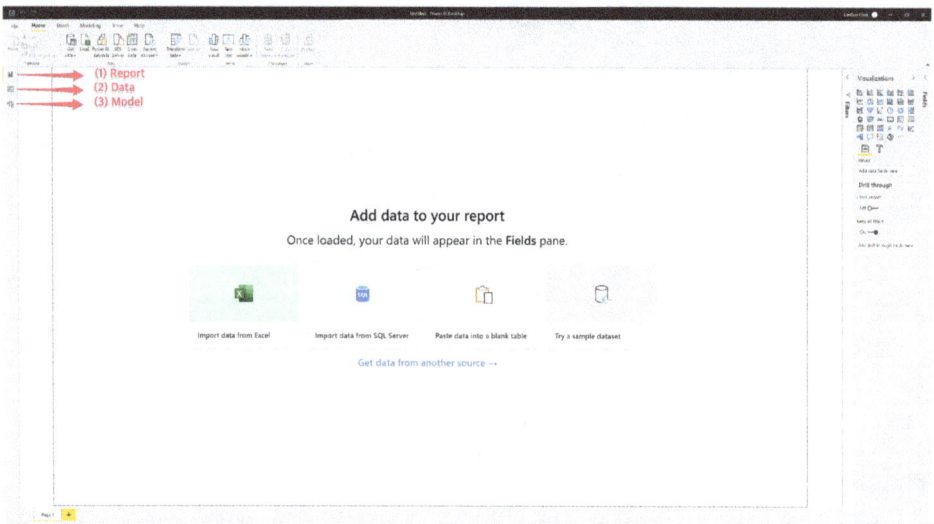

**Figure 11-6**    Configure the Power BI Desktop screen

(1) Report: Users can use the queries they create for great visualization with multiple pages and share them with others.

(2) Data: Users can view data in a report in a data model format that lets them add measures, create new columns, and manage relationships.

(3) Model: Users can obtain a graphical representation of the relationships that are set up in their data model and manage or modify as needed.

## 11.4  Data Import

Power BI Desktop lets users import a variety of data, including databases, web, online services, and Excel spreadsheets.

Click on "Get Data" from the menu and select "More…" Click to see the various data import methods.

### 11.4.1  *Import Open Data from Data World*

Find additional support for inserting data via the "Get Data" menu, as shown in Figure 11-7.

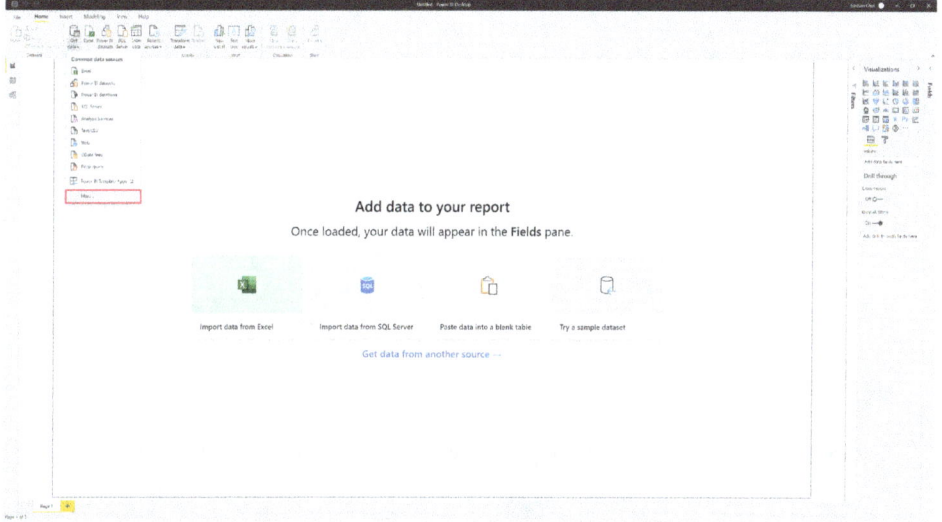

**Figure 11-7**   Additional data import method

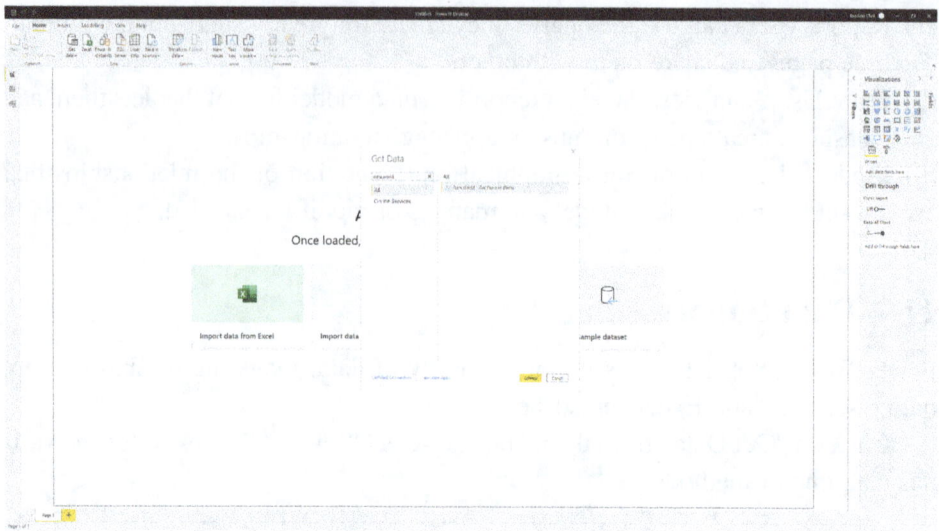

**Figure 11-8**    Data World

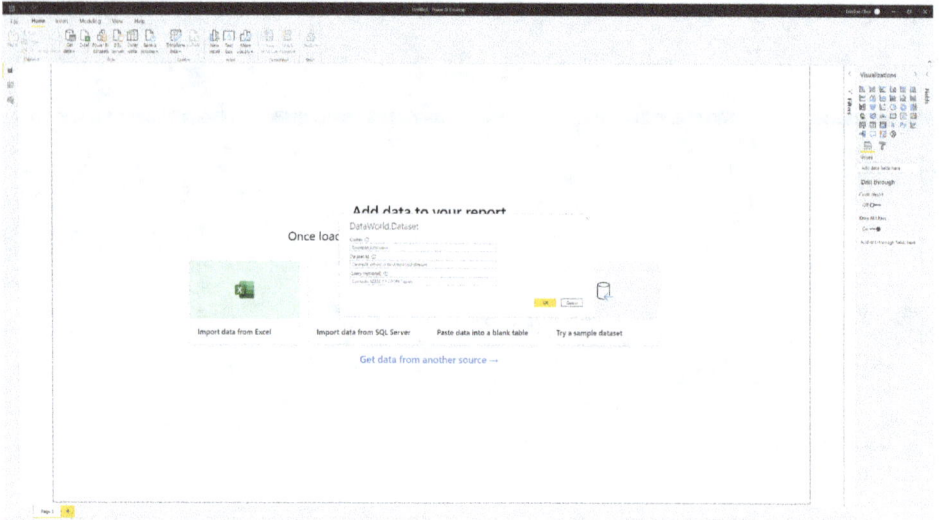

**Figure 11-9**    Parameter value input pop-up window for importing data from Data World

Search and click on "Data World" as shown in Figure 11-8.

In order to import data from Data World as shown in Figure 11-9, users must enter information in the "Owner", "Dataset Id", and "Query" fields.

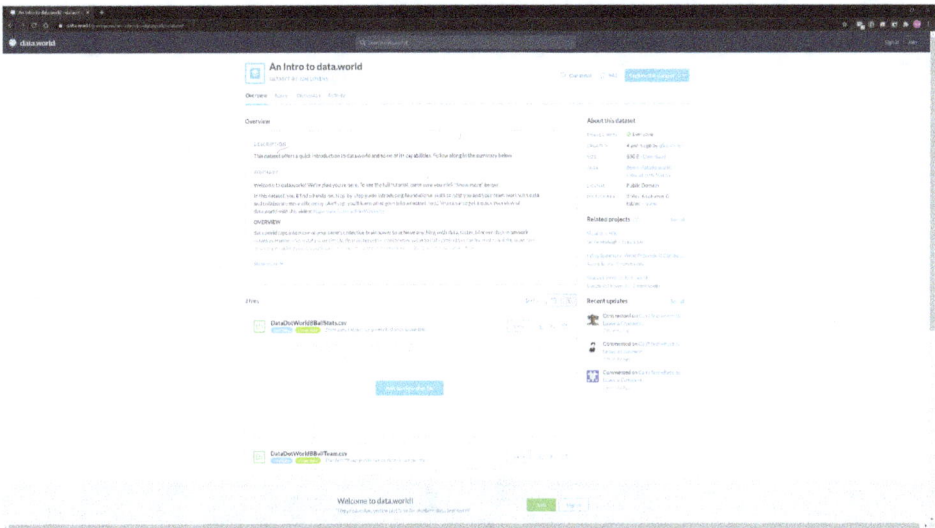

**Figure 11-10** Sample Site for Data Importing

The link in Figure 11-10 is as follows.

https://data.world/jonloyens/an-intro-to-dataworld-dataset

The reason for emphasizing the link is that the "Owner" and "Dataset Id" fields are essential to importing open data from Data World, as shown in the above address.

At the above link, "Jonloyens" is the owner, and "an-intro-to-dataworld-dataset" is the dataset Id.

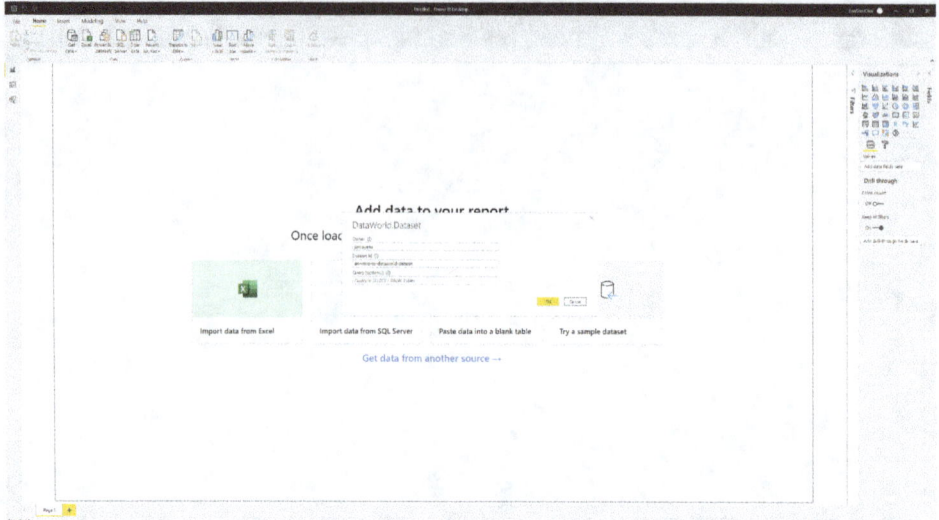

**Figure 11-11**  Insert parameter value to import data from Data World

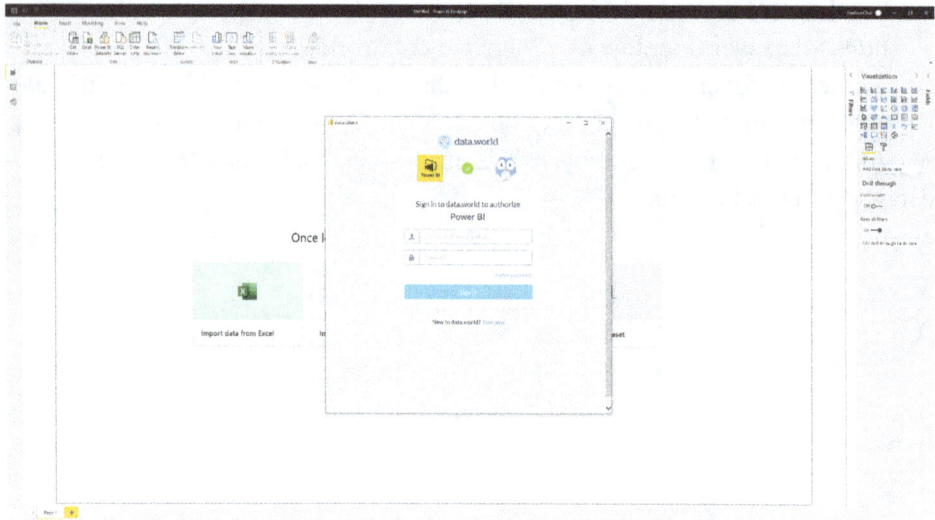

**Figure 11-12**  Login Pop-up window for Data World

Extract the "Owner" and "Dataset Id" of the above link and enter it in the parameters shown in Figure 11-11.

As shown in Figure 11-12, an account is required to import data from "Data World".

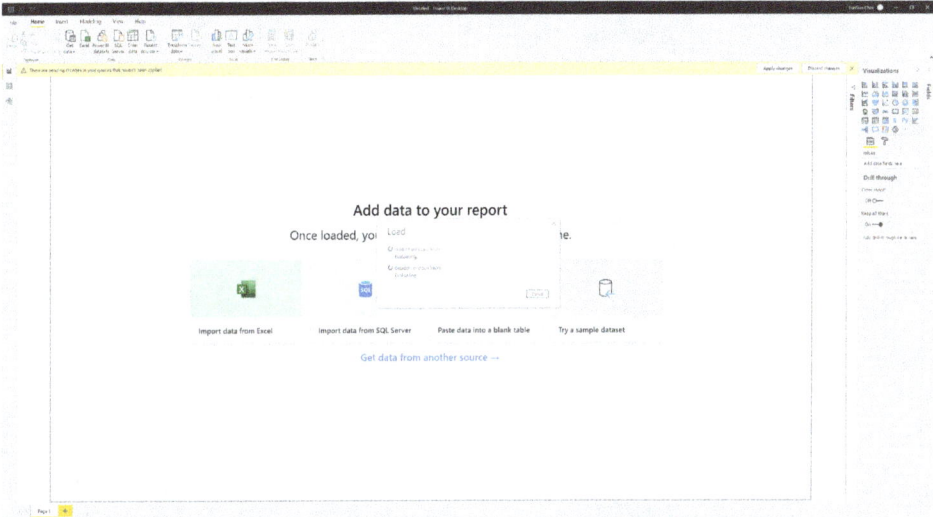

**Figure 11-13** Loading Pop-up window for importing data from Data World

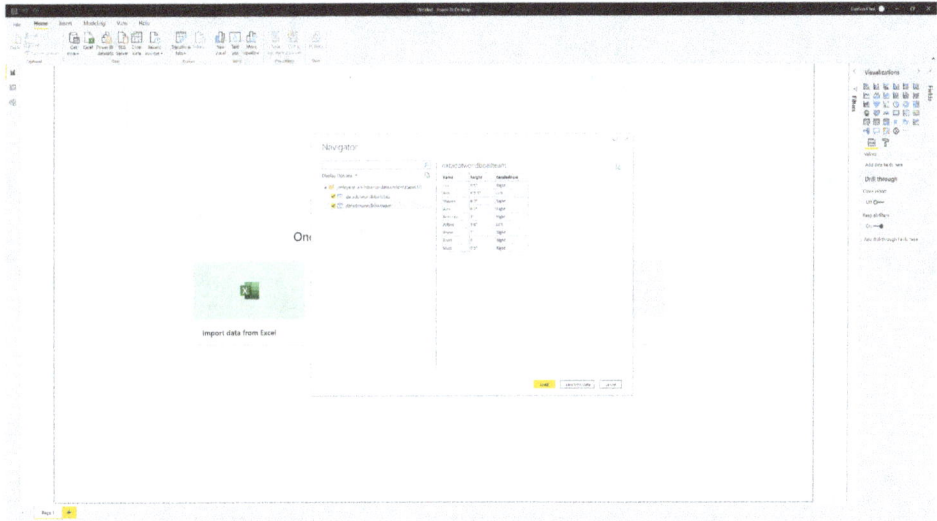

**Figure 11-14** Data World data importing success

When data importing is successful after login, the "Load" window appears, as shown in Figure 11-13.

After loading, you can check the data imported in Data World, as shown in Figure 11-14.

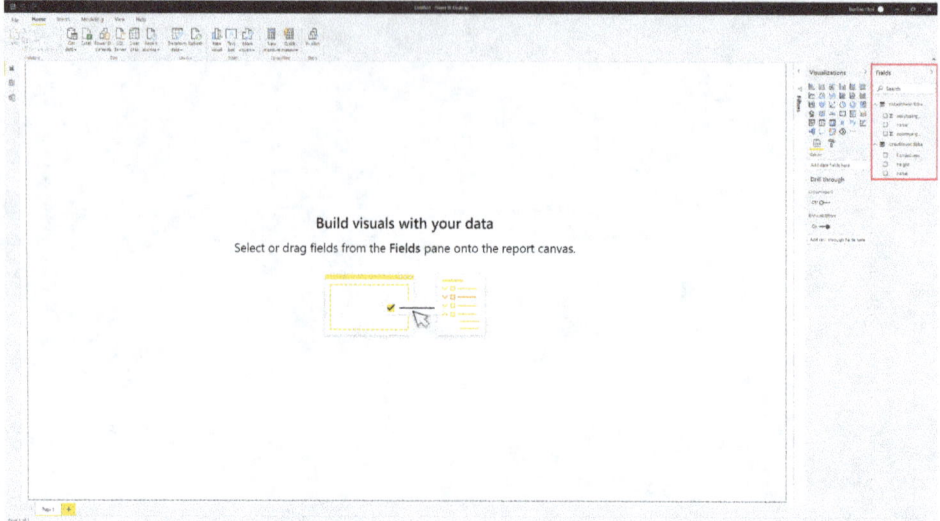

**Figure 11-15**    Data from "Data World" on Power BI

**Figure 11-16**    How "Data World" data on Power BI is utilized

After data importing is complete, data will appear in the "Fields" palette as shown in Figure 11-15.

The graph in Figure 11-16 shows that the data has been imported without error.

## 11.4.2 *Importing Excel File Data*

To import an Excel data file, click the "Excel" button in the "Get Data" menu, as shown in Figure 11-17. Excel file extensions supported are .xl, .xlsx, .xlsm, .xlsb, .xls, and .xlw files. Text and CSV files can be imported via the "Text/CSV" button at the bottom.

After loading, you can check the details of the data entered in the Excel file, as shown in Figure 11-18.

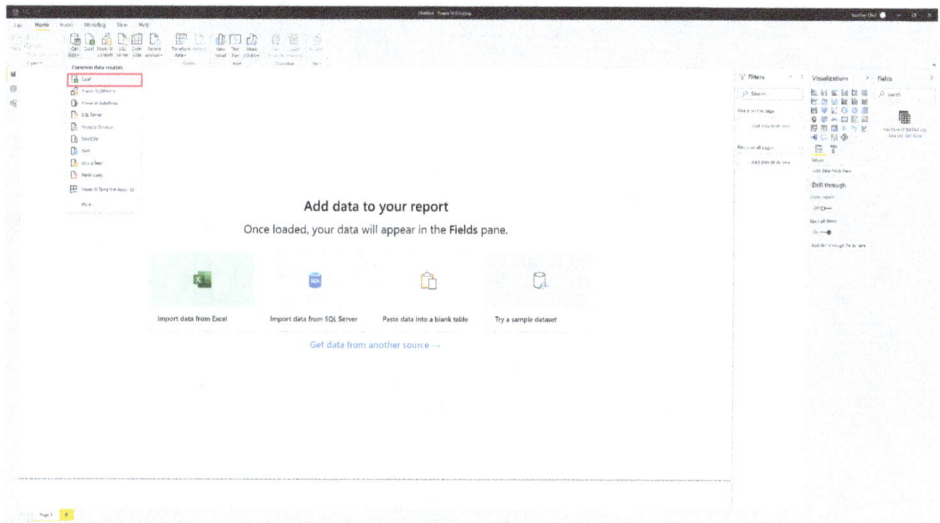

**Figure 11-17**   Data import method

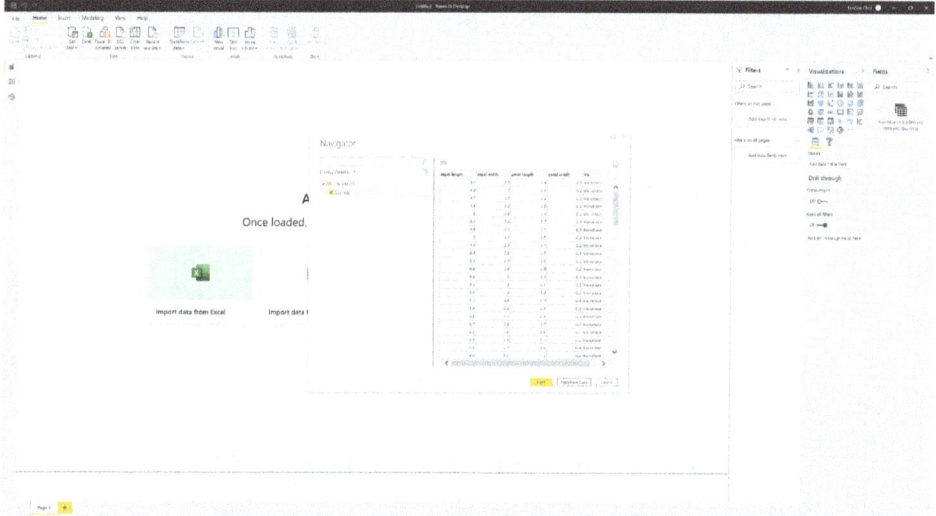

**Figure 11-18**   Excel Data Import Success

After data importing is complete, data will appear in the "Fields" palette, as shown in Figure 11-19.

The table in Figure 10-20 shows that the data has been imported without error.

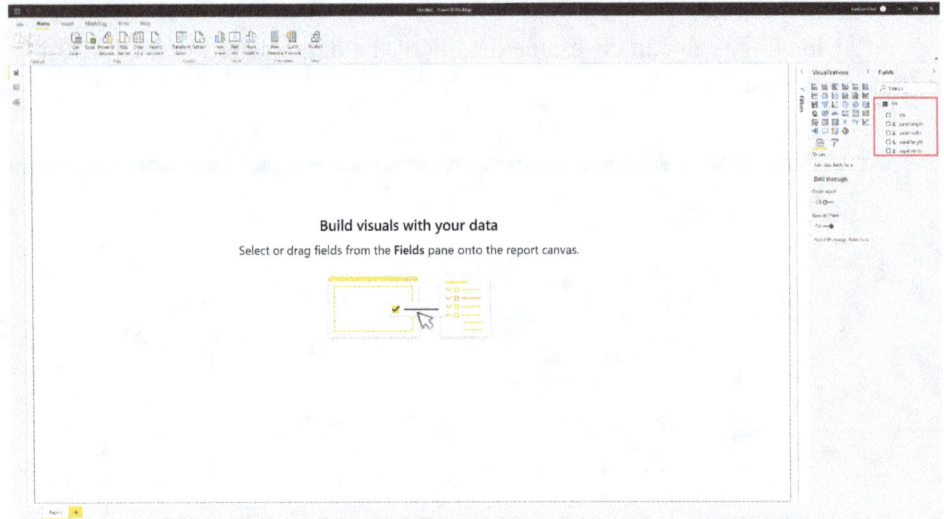

**Figure 11-19**  Excel Data on Power BI

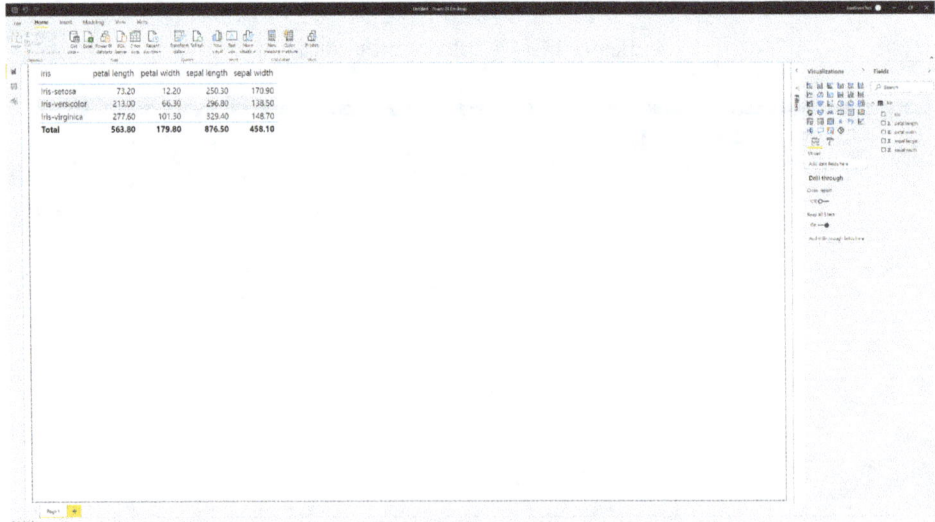

**Figure 11-20**  How "Excel" Data on Power BI is utilized

## 11.5  Introduction of Power BI Visualization Graph

To obtain meaningful patterns or business insight from the data set, visualization graphs suitable for the data set should be utilized. Explore the types and characteristics of the visualization graphs that Power BI provides natively.

### 11.5.1  *How to use Visualizations in Power BI*

You can conveniently visualize the inserted data through Power BI's "Visualizations" menu. The "Visualizations" menu is shown in the upper right corner of Figure 11-21.

Data importing should precede the application of visualization. Then, select the graph to apply from the visualization menu. Finalize the graph by selecting the data to be applied. The graph in the upper left-hand corner of Figure 11-22 can be freely adjusted by dragging and changing its size.

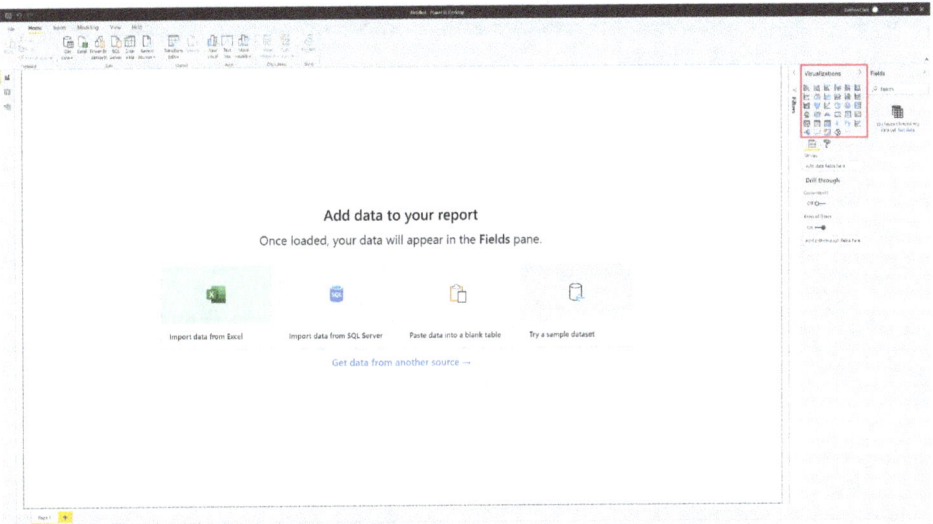

**Figure 11-21**   Visualization methods provided by Power BI

**Figure 11-22**   How to apply Power BI visualizations

**Figure 11-23**   Result of Power BI visualization

In addition, more detailed graphs can be easily created by adjusting the data to be entered into the "Axis", "Legend", and "Value" fields in Figure 11-23.

## 11.5.2 *Types of Visualization Chart*

Figure 11-24 shows a stacked bar chart. A cumulative stacked bar chart is a chart with individual elements stacked horizontally, making it easier to identify the contribution of individual elements to the total.

Figure 11-25 shows a stacked column chart. A stacked column chart is a chart with individual elements stacked vertically, making it easy to identify the contribution of individual elements to the total.

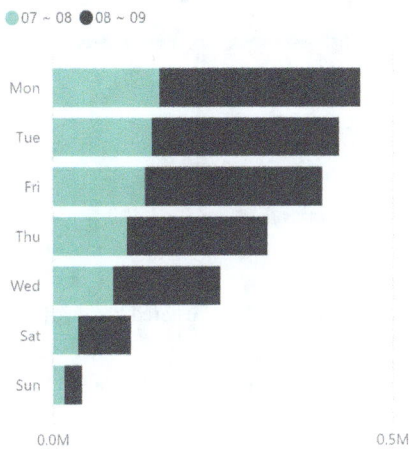

**Figure 11-24**　Stacked bar chart

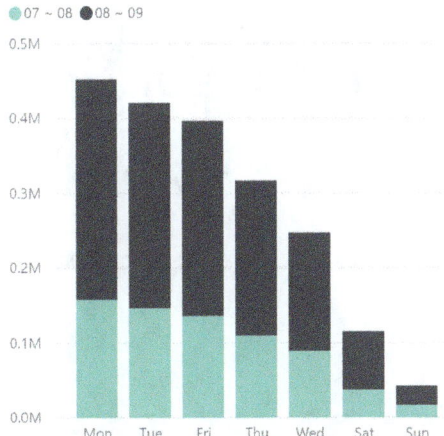

**Figure 11-25**　Stacked column chart

Figure 11-26 shows a clustered bar chart. A clustered bar chart compares individual data values by representing them as horizontal bars.

Figure 11-27 shows a clustered column chart. A clustered column chart compares the values of individual data by representing them as vertical bars.

**Figure 11-26**   Clustered bar chart

**Figure 11-27**   Clustered column chart

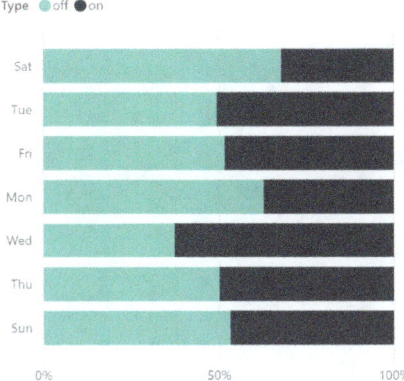

**Figure 11-28**　100% Stacked bar chart

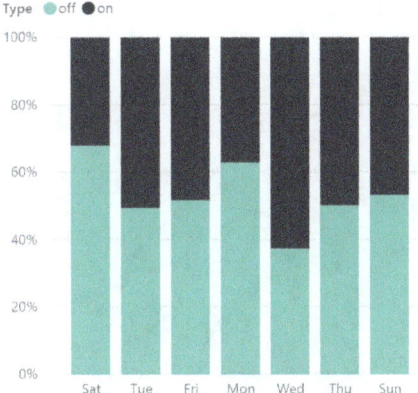

**Figure 11-29**　100% Stacked column chart

Figure 11-28 shows a 100% stacked bar chart. A 100% stacked bar chart converts the sum of individual elements into 100%, representing the weight of individual data in horizontal bars.

Figure 11-29 shows a 100% stacked column chart. A 100% stacked column chart converts the sum of individual elements into 100%, representing the weight of individual data in vertical bars.

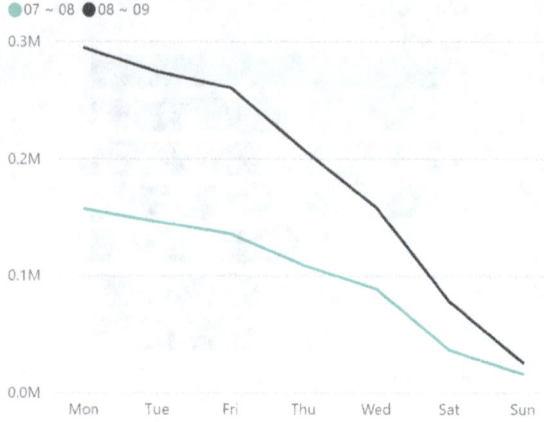

**Figure 11-30**  Line chart

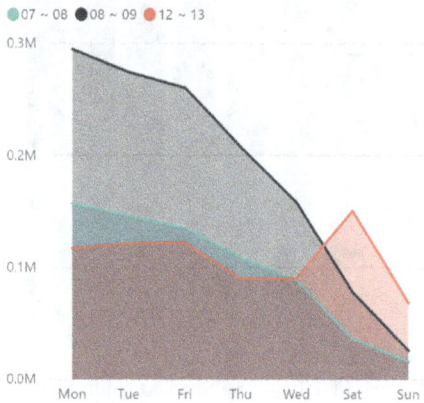

**Figure 11-31**  Area chart

Figure 11-30 shows a line chart. A line chart is suitable for determining trends in data over time.

Figure 11-31 shows an area chart. An area chart makes it easy to identify the magnitude of data changes over time.

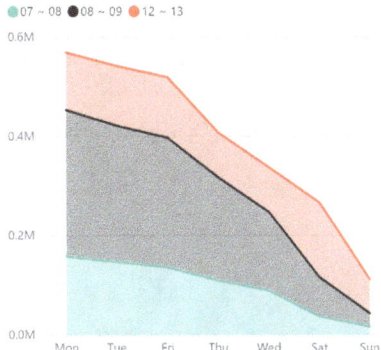

**Figure 11-32**  Stacked area chart

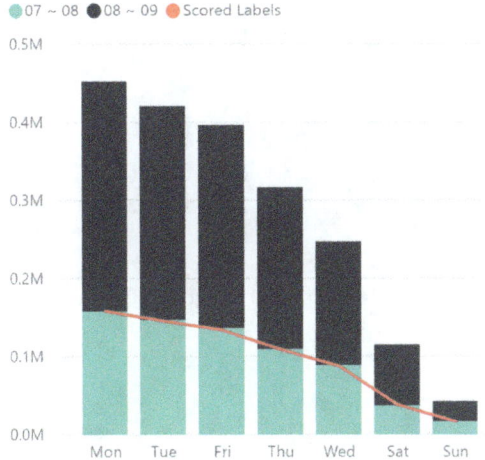

**Figure 11-33**  Line and Stacked column chart

Figure 11-32 shows a stacked area chart. A stacked area chart makes it easy to identify trends and contributions between data.

Figure 11-33 shows a line and stacked column chart. A line and stacked column chart is used to facilitate understanding of data when data values are broad.

Figure 11-34 shows a line and clustered column chart. A line and clustered column chart makes it easy to understand data if there is a wide range of data values.

Figure 11-35 shows a ribbon chart. A ribbon chart can quickly visualize data with the highest rank.

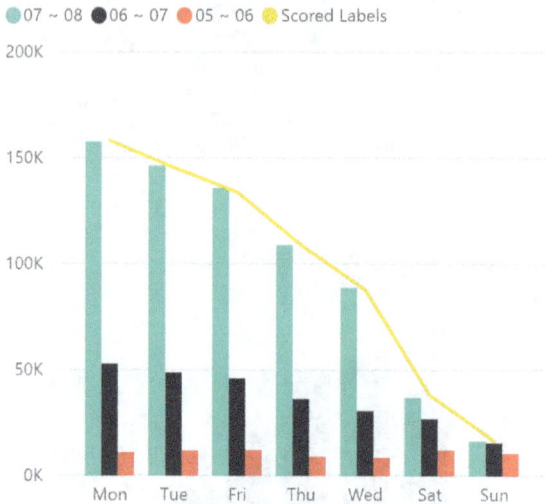

**Figure 11-34**   Line and Clustered column chart

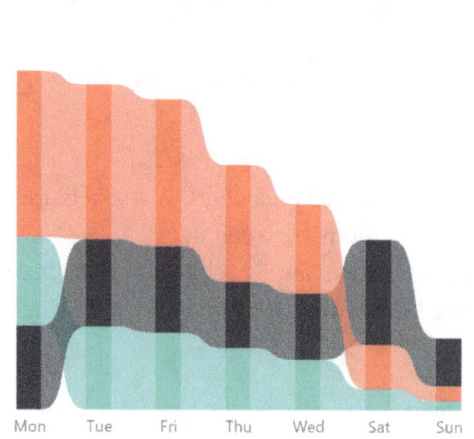

**Figure 11-35**   Ribbon chart

Figure 11-36 shows a waterfall chart. A waterfall chart is a comparison of how cumulative values are affected by adding or subtracting values. In particular, this is used when there are changes in measurements over time series or other categories, and when significant changes affecting the total value are audited.

Figure 11-37 shows a scatter chart. A scatter chart can find a correlation between two numbers. It is especially useful if you wish to display the relationship between two numeric values and if you wish to draw two numerical groups in x, y coordinates.

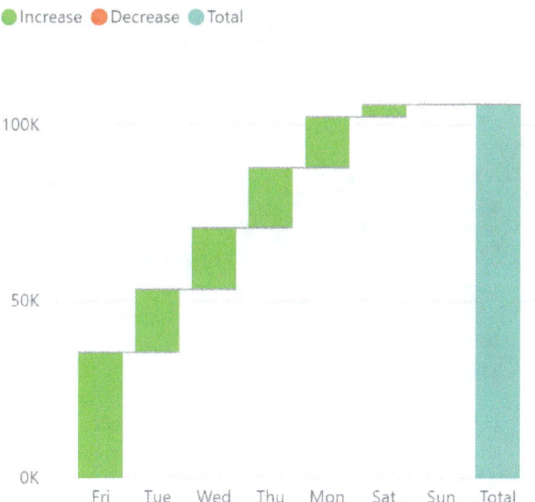

**Figure 11-36**   Waterfall chart

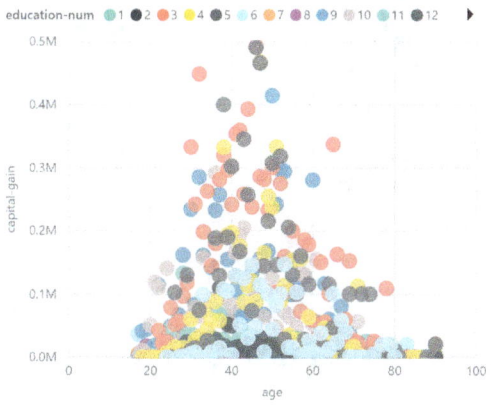

**Figure 11-37**   Scatter chart

Figures 11-38 shows a pie chart. A pie chart represents the percentage of individual elements compared to the whole.

Figure 11-39 shows a donut chart. A donut chart is similar to a pie chart in that it represents the relationship of parts to the whole. The difference is that the center is empty, and there is room for labels or icons.

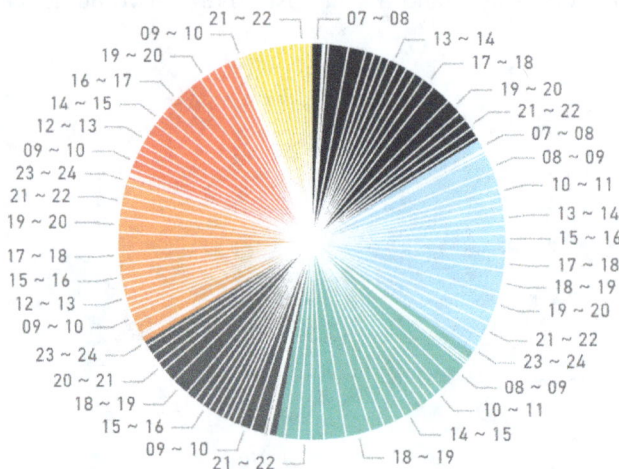

**Figure 11-38**   Pie chart

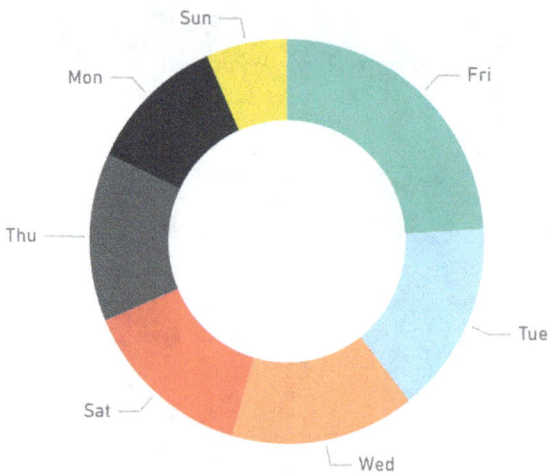

**Figure 11-39**   Donut chart

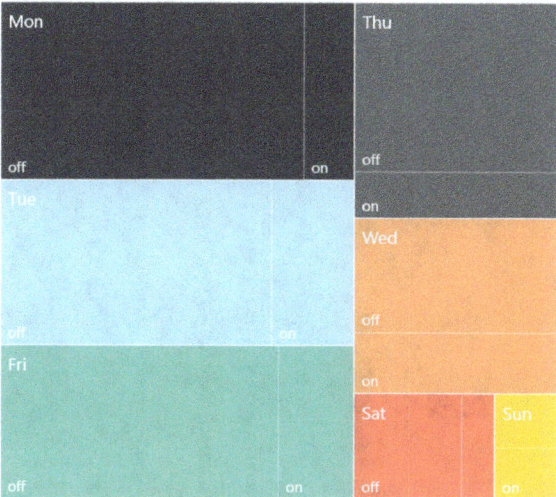

**Figure 11-40**   Treemap

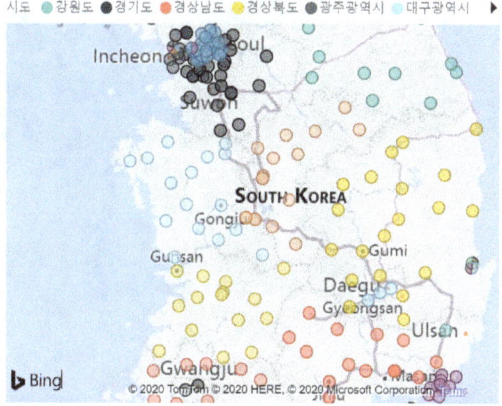

**Figure 11-41**   Map

Figure 11-40 shows a treemap. A treemap displays hierarchical data as a set of nested squares. The rectangles are arranged in order of size, with the largest being in the upper left, and the smallest being in the lower right. In particular, it is used to represent large quantities of hierarchical data or to indicate the ratio between each part and the whole.

Figure 11-41 shows a map. A map links location data to the map to compare the size of quantitative data.

**Figure 11-42** Filled map

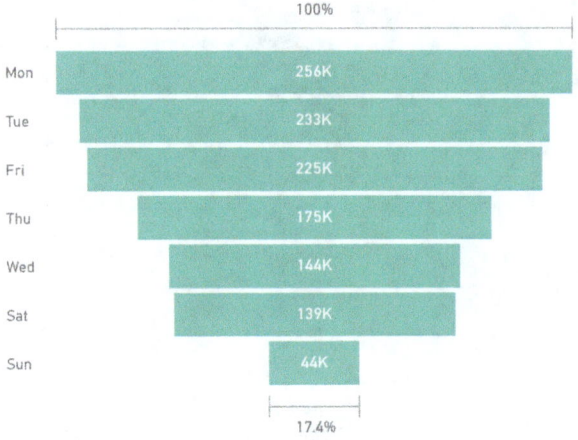

**Figure 11-43** Funnel chart

Figure 11-42 shows a filled map. A filled map uses shading, hues, and patterns to indicate how a particular value differs based on a particular place or region. It is used mainly when quantitative information is displayed on maps and when working with socio-economic data.

Figure 11-43 shows a funnel chart. A funnel chart is used to visualize sequentially connected processes, i.e. linear processes.

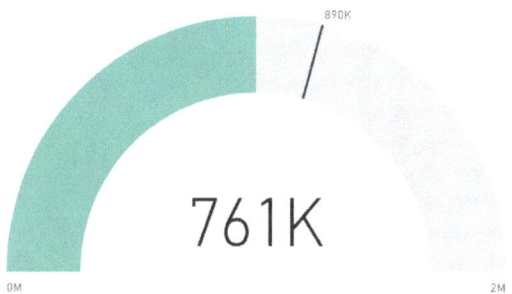

890K

761K

0M                                                              2M

**Figure 11-44**   Gauge

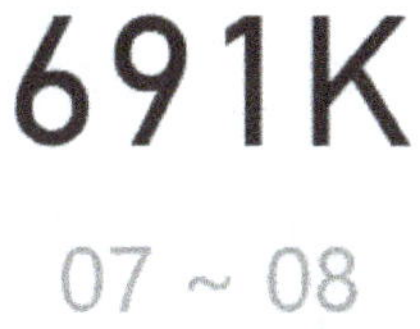

# 691K

07 ~ 08

**Figure 11-45**   Card

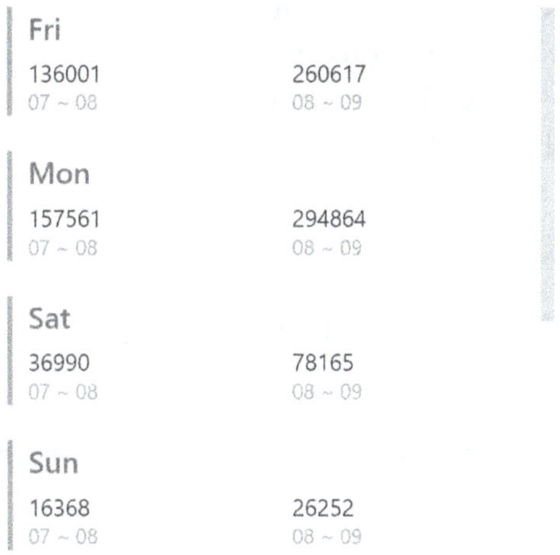

Fri
136001                    260617
07 ~ 08                   08 ~ 09

Mon
157561                    294864
07 ~ 08                   08 ~ 09

Sat
36990                     78165
07 ~ 08                   08 ~ 09

Sun
16368                     26252
07 ~ 08                   08 ~ 09

**Figure 11-46**   Multi-Row card

Figure 11-44 shows a gauge. A gauge displays a single value for measuring progress or KPI for the target. The line of the chart represents the target value, and the shade represents the progress towards the goal.

Figure 11-45 shows a card. A card shows only one element on a tile in large numbers. It is used mainly to track only the most essential numbers.

Figures 11-46 show a multi-row card. A multi-row card shows a large number of values for several elements on a tile.

**Figure 11-47**   KPI

Figure 11-47 shows a KPI. A KPI is a graph showing the degree of progress for measurable goals.

## 11.6  Using Learning Results with Azure Machine Learning Studio

Although artificial intelligence with high predictability is essential, it is also essential to adequately express learning results through visualization. Import data into Power BI and try to visualize the learning results of Azure Machine Learning Studio using an Excel file and R.

### 11.6.1 *Excel File*

Place and connect the "Convert to CSV" block, as shown in Figure 11-48 to extract the CSV file.

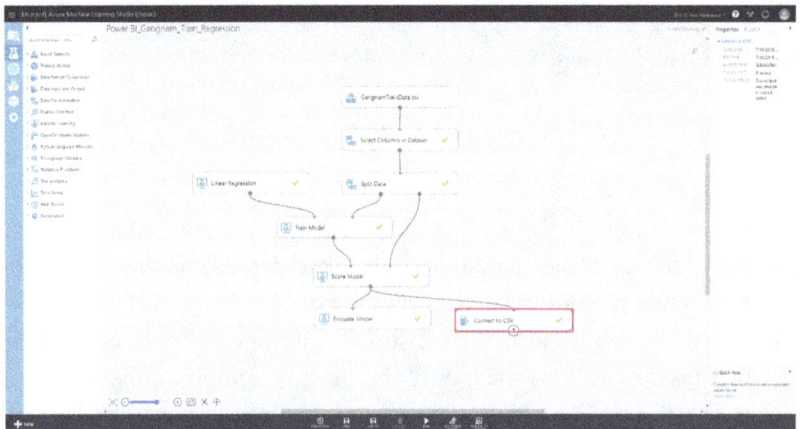

**Figure 11-48**   Arrangement to obtain learning result data

Click the "Run" button to learn. Then, as shown in Figure 11-49, right-click, and download the CSV file.

Click the "Excel" button in the "Get Data" menu to insert the downloaded CSV file, as shown in Figure 11-50. The CSV file can be imported via the "Text/CSV" button at the bottom, however the CSV file can also be imported through "All files (*.*)" when selecting the file.

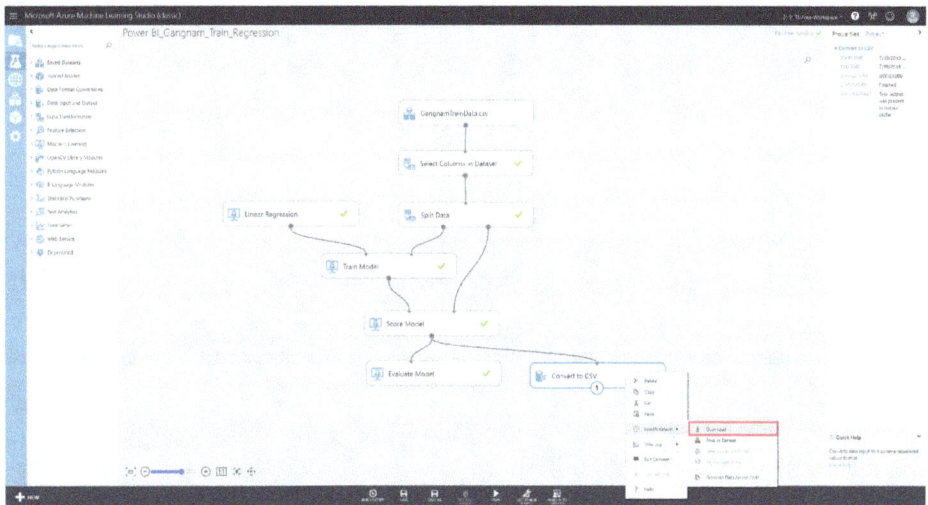

**Figure 11-49**  Download Excel data

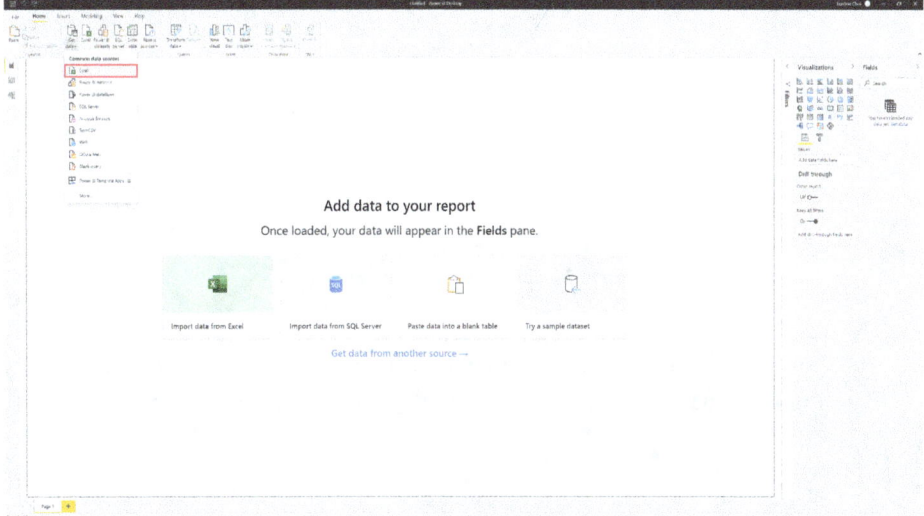

**Figure 11-50**  Data import method

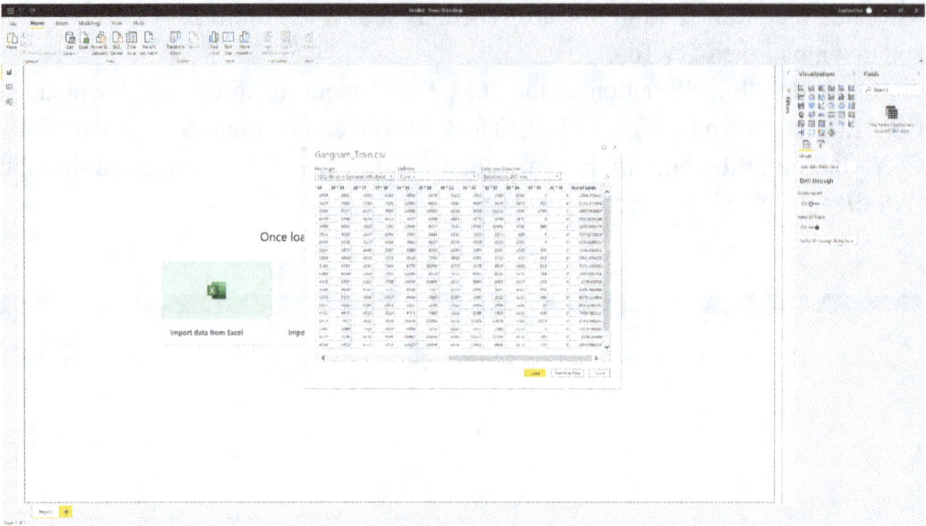

**Figure 11-51**    Excel data import success

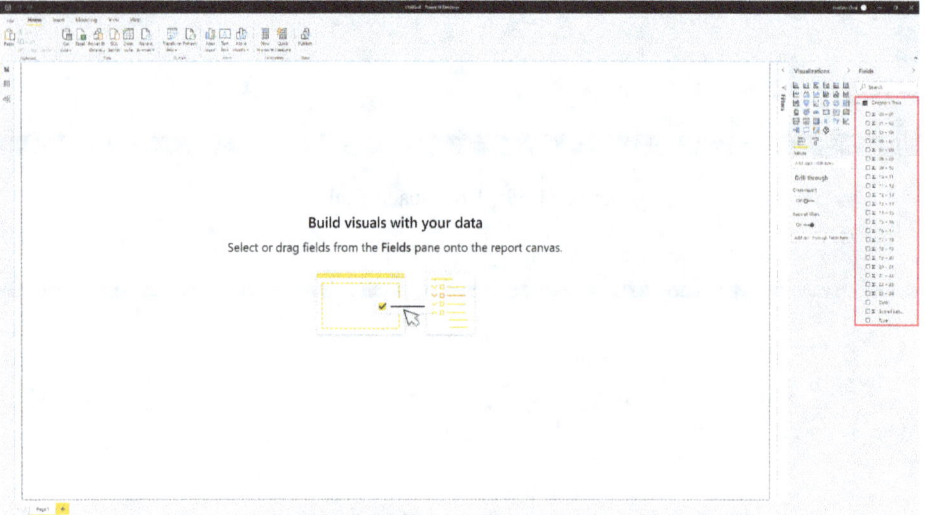

**Figure 11-52**    Excel data on Power BI

As shown in Figure 11-51, Excel data has been successfully imported. The imported data can be found on the right side of Figure 11-52.

**Figure 11-53**   Clustered column chart

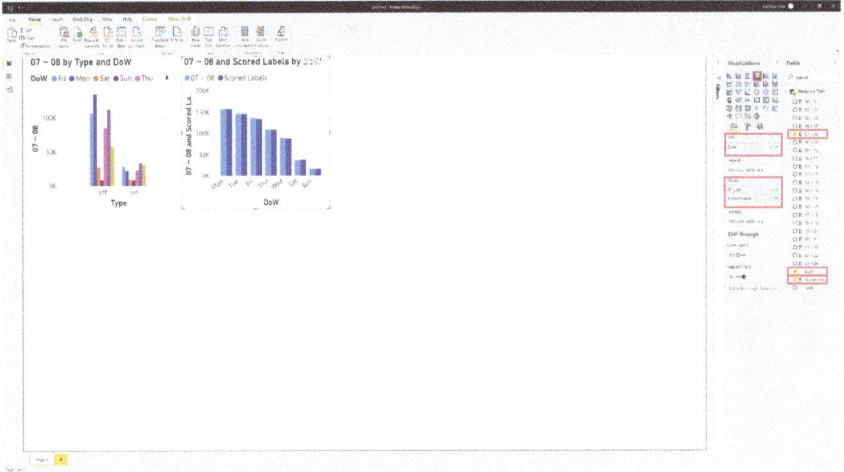

**Figure 11-54**   Clustered column chart

Figure 11-53 shows the imported data being applied to a clustered column chart. You can compare the number of arrivals and departures according to the day of week. The results show that there are a lot more people getting off than on.

Figure 11-54 shows a comparison of the number people getting on or off according to the day of week and the values predicted by Azure Machine Learning Studio in the form of a clustered column chart. As shown above, data inserted in conjunction with Azure Machine Learning Studio have scored labels, which allows for the comparison of actual and predicted values.

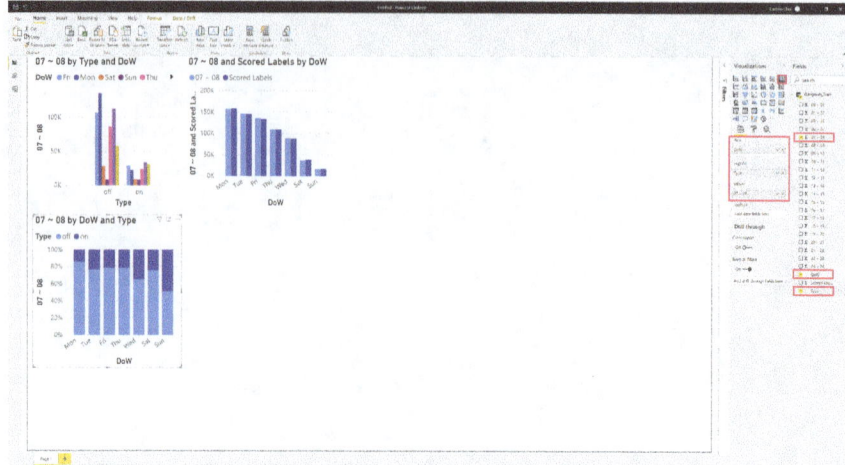

**Figure 11-55**   100% Stacked column chart

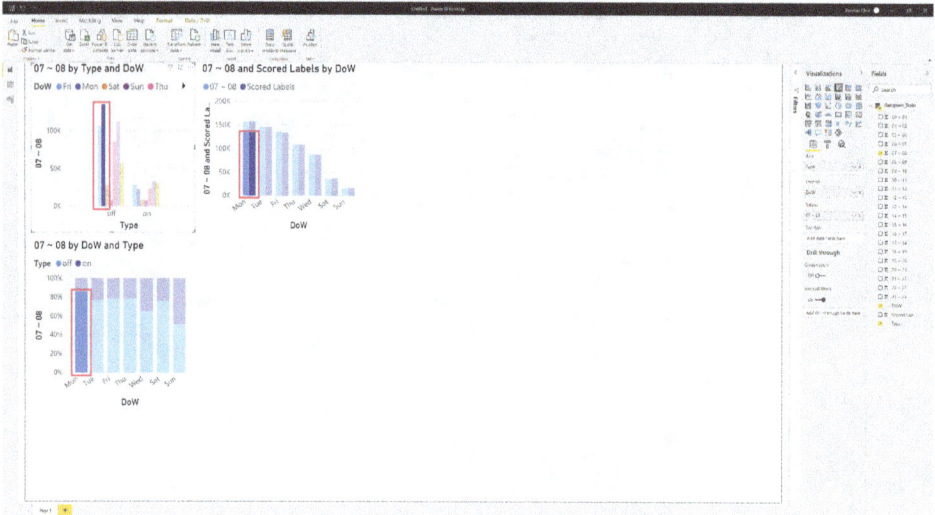

**Figure 11-56**   View selected items

Figure 11-55 shows the results applied to a 100% stacked column chart. You can see that the number of people getting on or off on Sunday is relatively unbiased based on the day of the week.

The complete graph shown in Figure 11-56 shows the selected item in real-time.

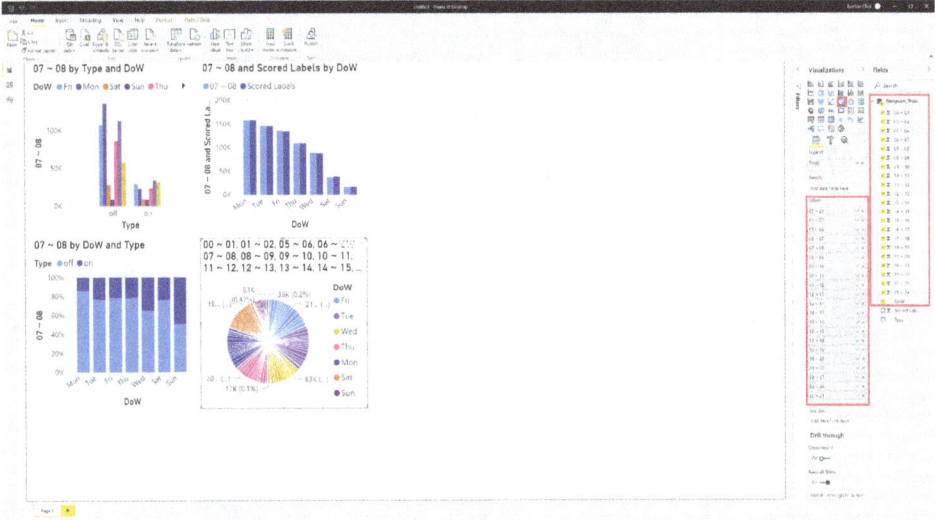

**Figure 11-57**    Pie chart

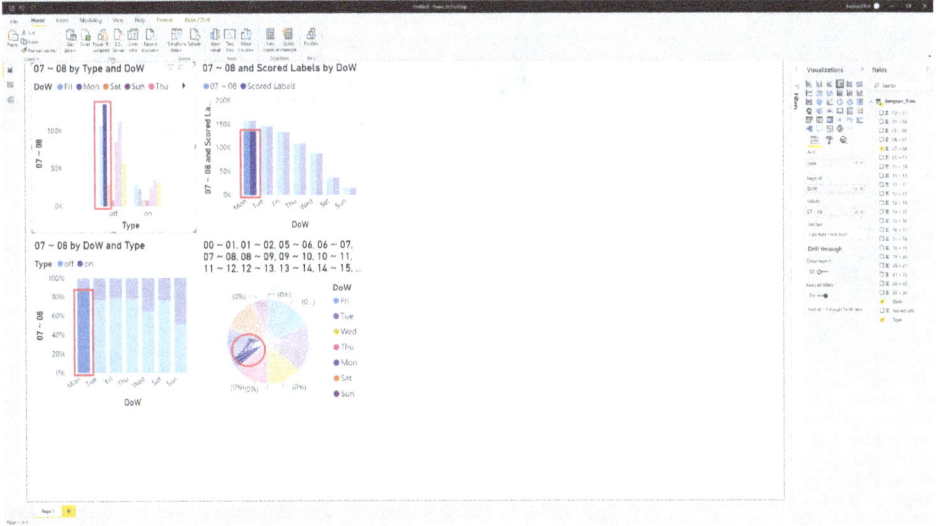

**Figure 11-58**    View selected items

As shown in Figure 11-57, we can use a pie chart to see the ratio of the overall configuration data.

Also, the pie chart can also visualize the selected items in real-time, as shown in Figure 11-58.

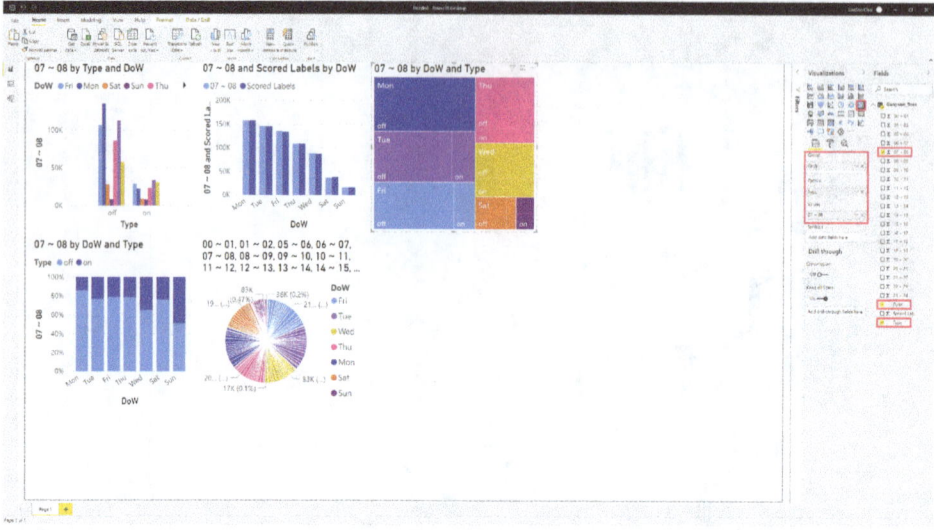

**Figure 11-59**    Treemap

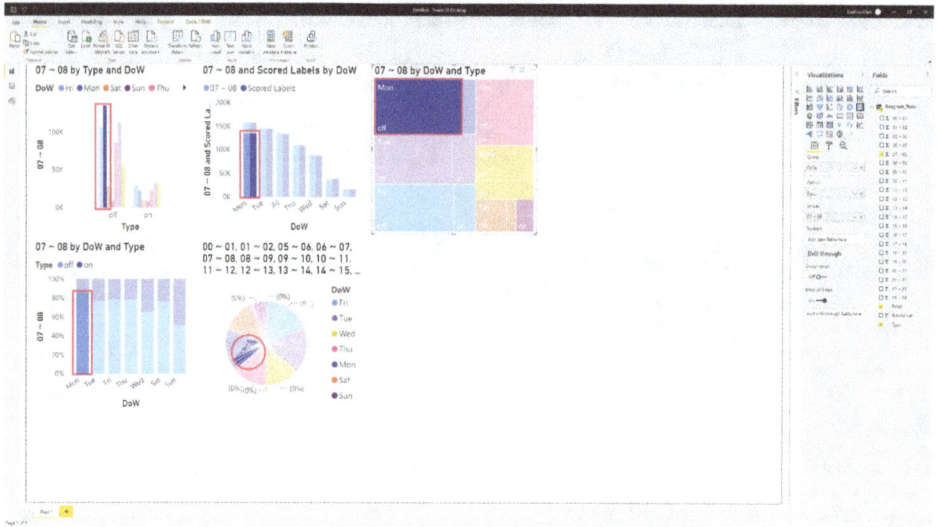

**Figure 11-60**    View selected items

Figure 11-59 shows a treemap. The rate of people getting on or off each day can be seen as a rectangular area.

A "treemap" allows you to view selected items in real-time, as shown in Figure 11-60. Because of its intuitive appearance, it is sometimes used as a button for data selection.

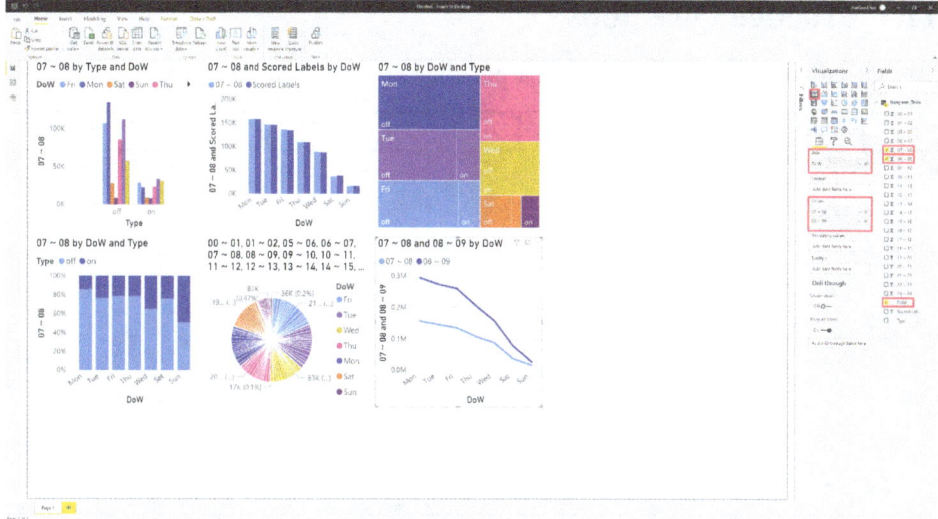

**Figure 11-61**   Line chart

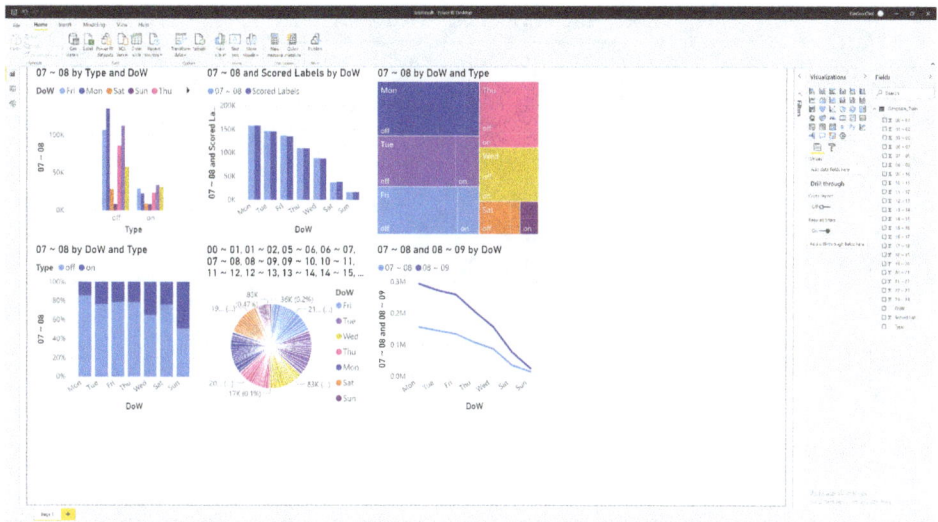

**Figure 11-62**   Visualization results

Figure 11-61 shows a line chart. A noticeable downward trend can be seen from Monday to Sunday.

Figure 11-62 shows the application results. In conjunction with Azure Machine Learning Studio, data can be imported, and insight can be gleaned visually and through prediction results.

## 11.7 Practice Questions

Q1. Visualize the following data in a pie chart. Please write in the ratio.

| Data | 0 | 1 | 0 | 2 | 4 | 3 | 0 | 0 | 2 | 1 | 3 | 4 | 4 | 2 | 2 | 1 | 0 | 1 | 4 | 3 |
|------|---|---|---|---|---|---|---|---|---|---|---|---|---|---|---|---|---|---|---|---|

Q2. Visualize the following data in a 100% Stacked Column Chart

| Mon | Tue | Fri | Thu | Wed | Sat | Sun |
|-----|-----|-----|-----|-----|-----|-----|
| 0 | 0 | 1 | 1 | 0 | 0 | 1 |
| 0 | 0 | 1 | 1 | 1 | 0 | 0 |
| 0 | 1 | 1 | 0 | 1 | 0 | 1 |
| 0 | 0 | 1 | 1 | 1 | 1 | 0 |
| 1 | 0 | 1 | 0 | 1 | 1 | 1 |
| 1 | 1 | 1 | 1 | 0 | 1 | 0 |
| 0 | 1 | 0 | 0 | 0 | 1 | 0 |
| 1 | 0 | 1 | 1 | 0 | 1 | 1 |
| 1 | 0 | 0 | 1 | 0 | 1 | 1 |
| 1 | 1 | 1 | 0 | 0 | 1 | 0 |

# Chapter 12

# Visualization with R in Power BI

## 12.1 Introduction of R

If users proceed to deploy Azure Machine Learning Studio web service with R Studio, they will be able to proceed without any problem. However, there are some limitations when using R's visuals with Power BI Desktop:

(1) Data size limitations—data used by the R visual for plotting is limited to 150,000 rows. If more than 150,000 rows are selected, only the top 150,000 rows are used and a message is displayed on the image.
(2) Calculation time limitation—if an R visual calculation exceeds five minutes the execution times out, resulting in an error.
(3) Relationships—as with other Power BI Desktop visuals, if data fields from different tables with no defined relationship between them are selected, an error occurs.
(4) R visuals are refreshed upon data updates, data filtering, and data highlighting. However, the image itself is not interactive and cannot be the source of cross-filtering.
(5) R visuals respond when you highlight other visuals, but you cannot click on elements in the R visual in order to cross-filter other elements.
(6) Only plots that are plotted to the R default display device are displayed correctly on the canvas. Avoid explicitly using a different R display device.
(7) In this release, RRO installations are not automatically identified by the 32-bit version of Power BI Desktop, so you must manually provide the path to the R installation directory in **Options and settings > Options > R Scripting**.

## 12.2 How to use the R Script Editor

Click "R Script" in the "Visualization" category on the right side of Figure 12-1. The R Script editor will appear, as shown in Figure 12-2.

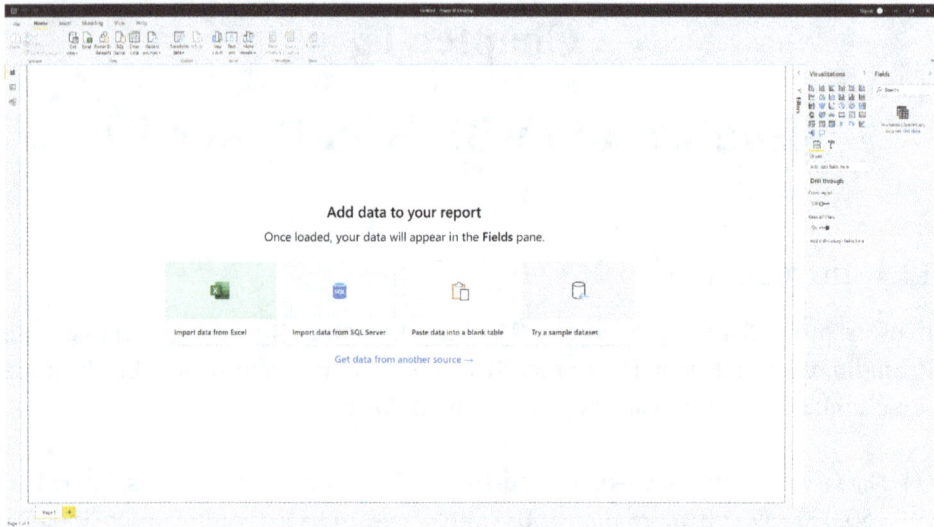

**Figure 12-1**   R Script Editor location

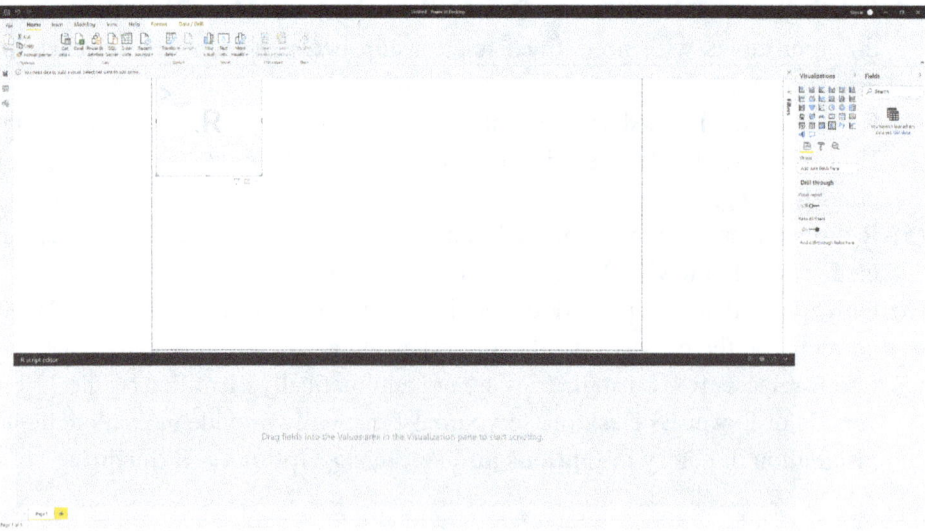

**Figure 12-2**   R Script editor

Import the data, as shown in Figure 12-3 and click on the data you want to use in the R Script editor. If you check the data to use on the right, you can see that the script is automatically modified.

You can run the modified R Script by clicking on the box in Figure 12-4.

**Figure 12-3**   Import data in R Script editor

**Figure 12-4**   How to run R Script

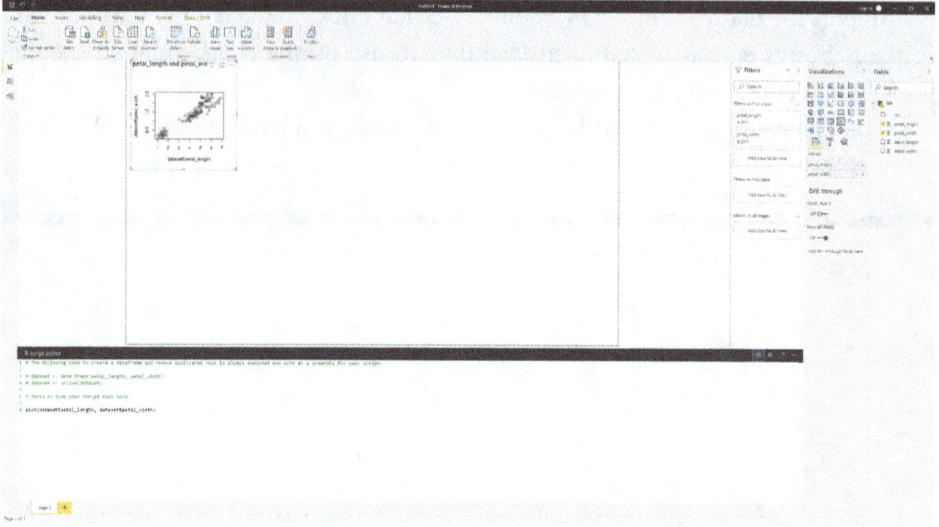

**Figure 12-5**   R Script execution result

You can see the R Script executed in Figure 12-5.

## 12.3  Visualization for Data Analysis Using Power BI R Script

Before applying visualizations for data analysis, let's learn about variables. Variables are divided into numerical data and categorical data according to their characteristics.

Numerical data such as length, power, and quantity is represented numerically, and category data such as gender and region is represented at predetermined levels.

Numerical data has numerical values of attributes that are quantitative, and category data has levels of attributes that are qualitative.

Depending on the number of variables, the plot can be different.

Univariate data is data having one variable, and multivariate data means data composed of several variables.

Let's take advantage of the Power BI R Script function to visualize data analysis according to the characteristics of the data.

## 12.3.1 *Numerical Univariate Plot*

Find additional support for inserting data through the "Get Data" menu, as shown in Figure 12-6.

Search and click on the "R Script", as shown in Figure 12-7.

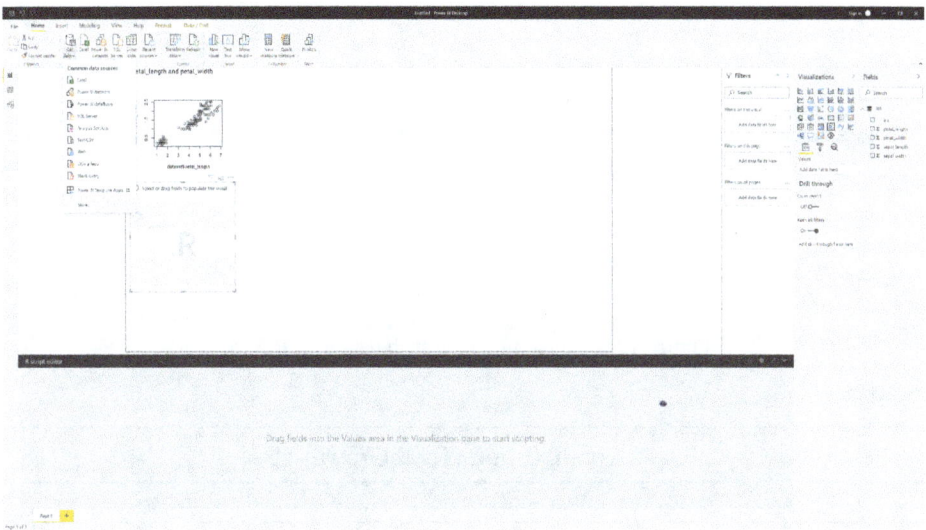

**Figure 12-6**   Additional data import method

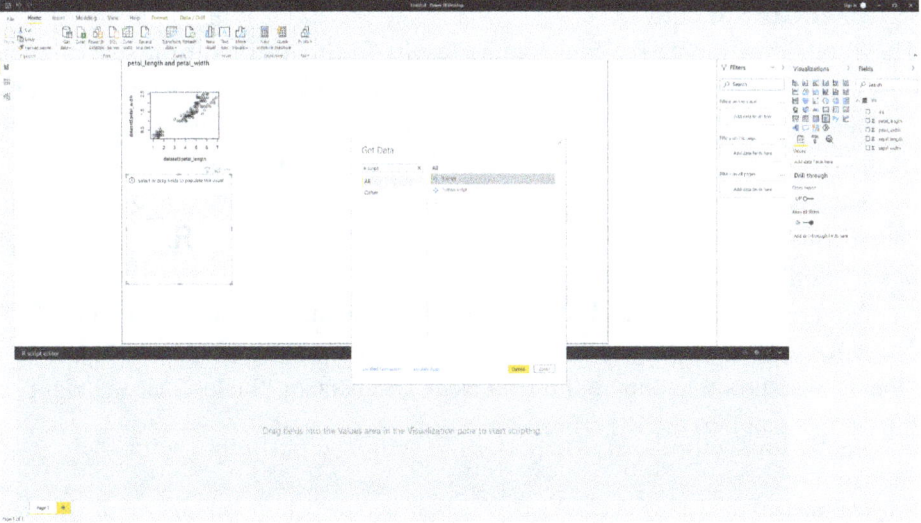

**Figure 12-7**   R Script data import method

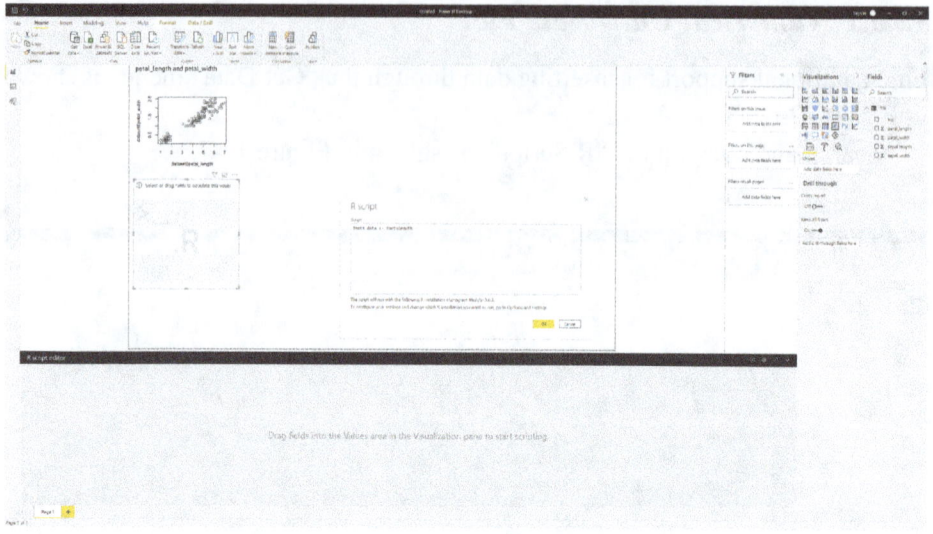

**Figure 12-8**    Import tooth growth data with R Script

**Source (12-8)**

```
Tooth_data <- ToothGrowth
```

Import the data supported by R to create the numerical univariate plot. Enter the source, as shown in Figure 12-8.

- **ToothGrowth data**

The ToothGrowth data set contains the results from an experiment studying the effects of vitamin C on tooth growth in 60 Guinea pigs. Each animal received one of three dose levels of vitamin C (0.5, 1, and 2 mg/day) by one of two delivery methods, orange juice or ascorbic acid (a form of vitamin C and coded as VC)).

The attributes of the ToothGrowth data are "len", meaning tooth length, "supp", meaning support type (VC or OJ), and "dose", meaning the number of doses in milligrams/day.

After loading, you can check the ToothGrowth data, as shown in Figure 12-9.

After data importing is complete, users can note that data now appears in "Fields" as shown in Figure 12-10. Check the data in "Fields" on the right to import the data into the R Script.

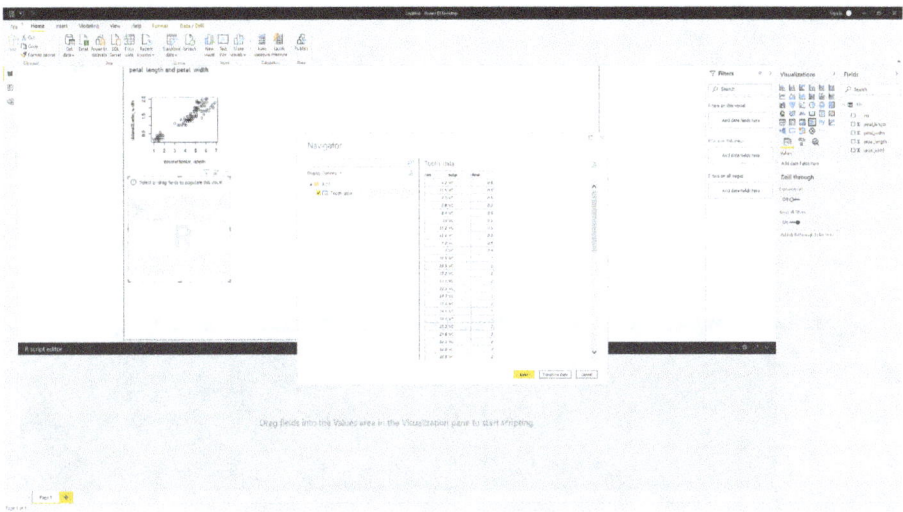

**Figure 12-9**   ToothGrowth data importing success

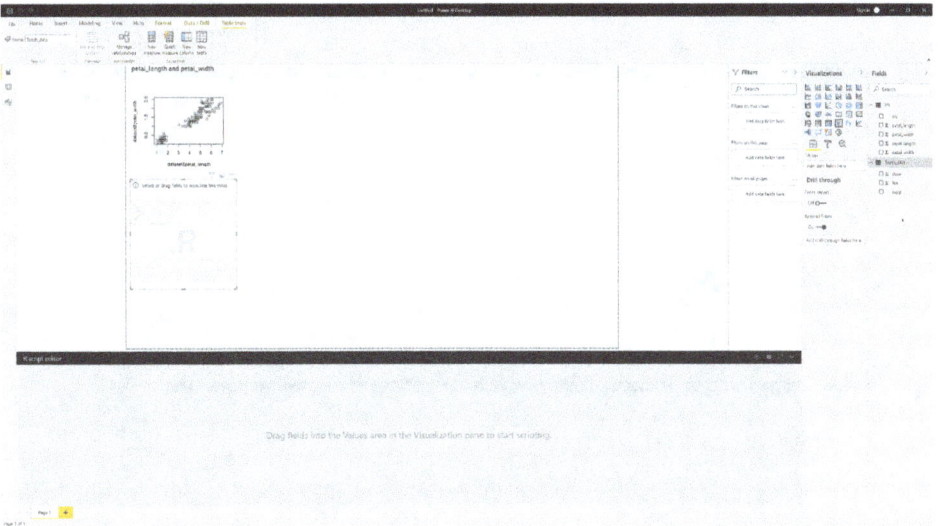

**Figure 12-10**   ToothGrowth data on Power BI

Numerical data such as "ToothGrowth" is continuous data. In general, data such as weight and height follow a normal distribution. Plot a histogram to see what distribution "ToothGrowth" follows. Enter the following source in the R Script window, as shown in Figure 12-11.

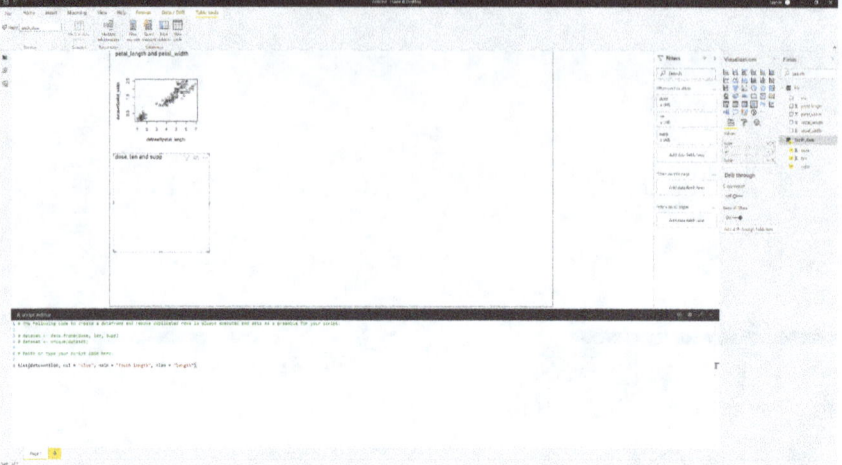

**Figure 12-11**　ToothGrowth histogram

Figure 12-12 shows a successful histogram having been drawn.

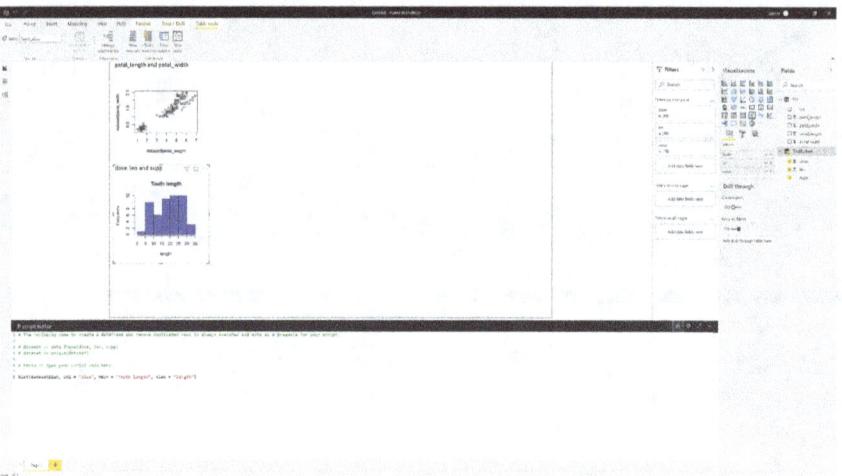

**Figure 12-12**　ToothGrowth histogram

## Source (12-12)

```
# dataset <- data.frame(len, supp, dose)
# dataset <- unique(dataset)

hist (dataset$len, col = "blue", main = "Tooth length", xlab = "length")
```

Now that we have a histogram let's analyze the "ToothGrowth data" in more detail using the distribution curve. Enter the following source in the R Script window, as shown in Figure 12-13.

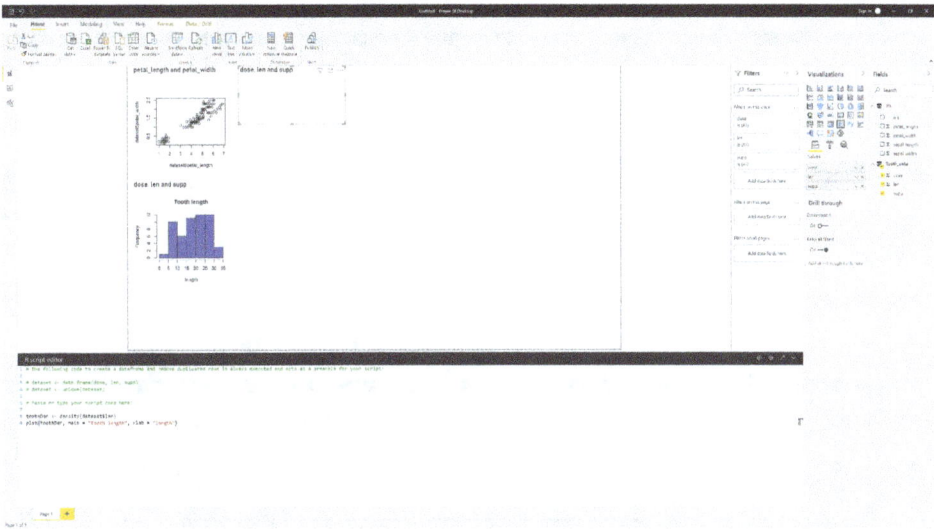

**Figure 12-13**  ToothGrowth distribution curve

You can see an error occur, as shown in Figure 12-14. This is an error due to the small range for the plot. Let's expand the scope of the R Script visualization.

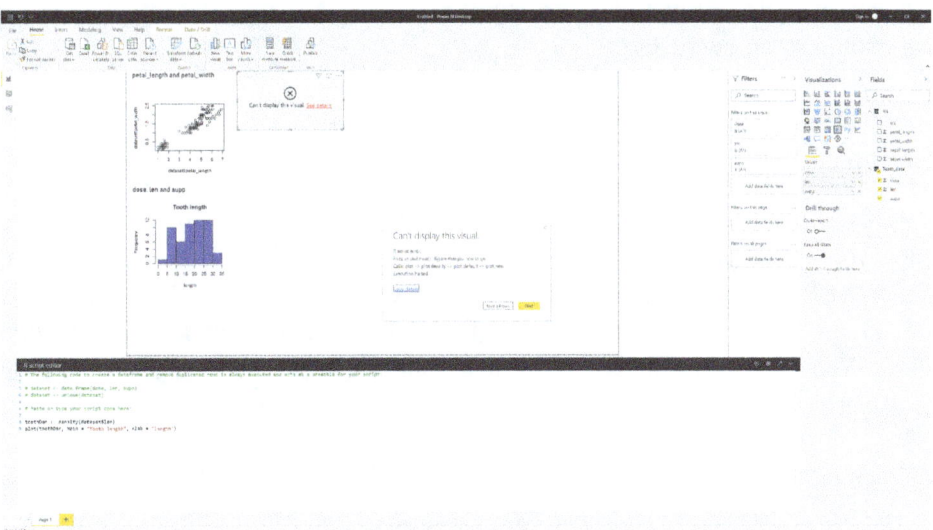

**Figure 12-14**  R Script error

By expanding the scope of the R Script visualization, you can see that it ran successfully. Figure 12-15 shows the distribution curve for the "ToothGrowth" data.

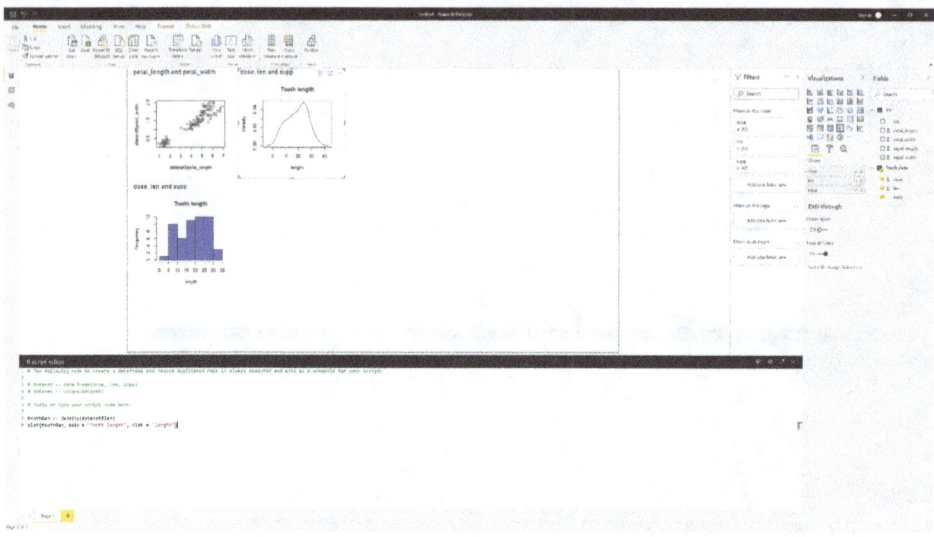

**Figure 12-15**   ToothGrowth distribution curve

## Source (12-15)

```
# dataset <- data.frame(dose, len, supp)
# dataset <- unique(dataset)

toothDen <- density(dataset$len)
plot(toothDen, main = "Tooth length", xlab ="length")
```

If you plot each histogram and distribution curve in the previous example as a plot, you can effectively determine the distribution of the "ToothGrowth" data. Enter the following source in the R Script window, as shown in Figure 12-16.

Figure 12-17 shows the result of plotting the histogram and the distribution curve. You can see that the "ToothGrowth" data does not exactly match the normal distribution.

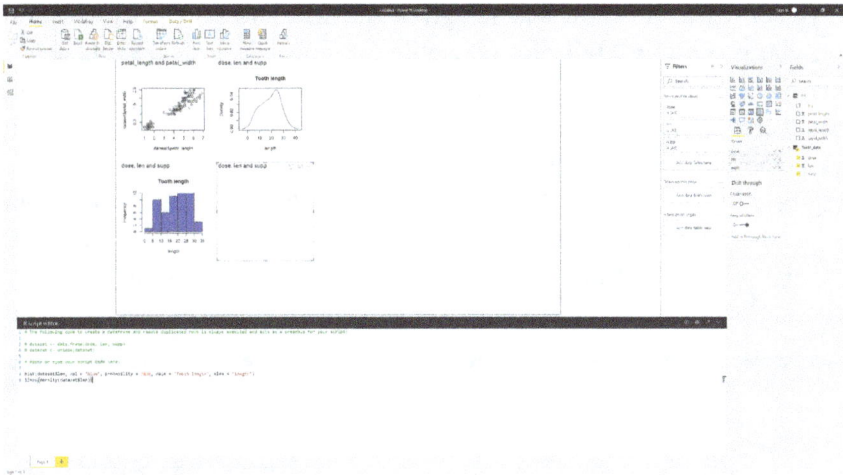

**Figure 12-16**  ToothGrowth histogram and distribution curve

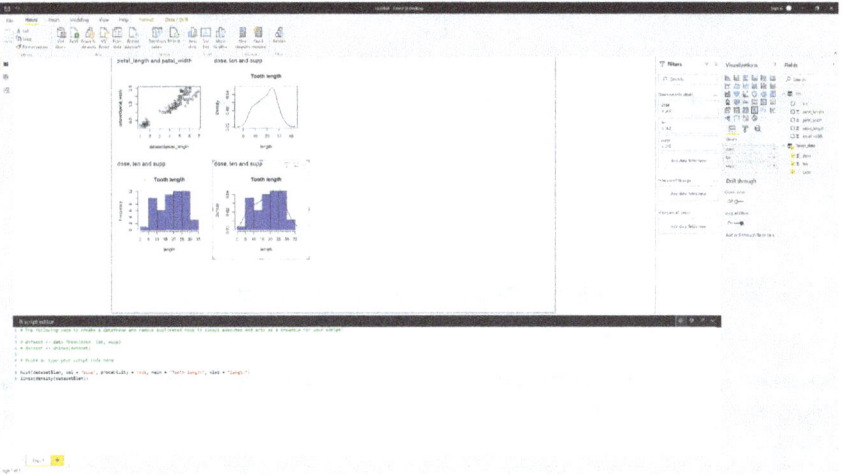

**Figure 12-17**  ToothGrowth histogram and distribution curve

## Source (12-17)

```
# dataset <- data.frame(dose, len, supp)
# dataset <- unique(dataset)

hist(dataset$len, col = "blue", probability = TRUE, main = "Tooth length", xlab
= "length")
lines(density(dataset$len))
```

If you only want to know whether given numerical data is normally distributed or not, you can plot a Q-Q plot (Quantile-Quantile plot). Q-Q plots determine whether given data follow a specific probability distribution. If the data follow a normal distribution, the points printed on the plot are straight. Enter the following source in the R Script window, as shown in Figure 12-18.

Figure 12-19 shows a Q-Q plot of "ToothGrowth" data. You can see that most of the points are on a straight line and follow a normal distribution, but the points at both ends show a distribution away from the normal distribution.

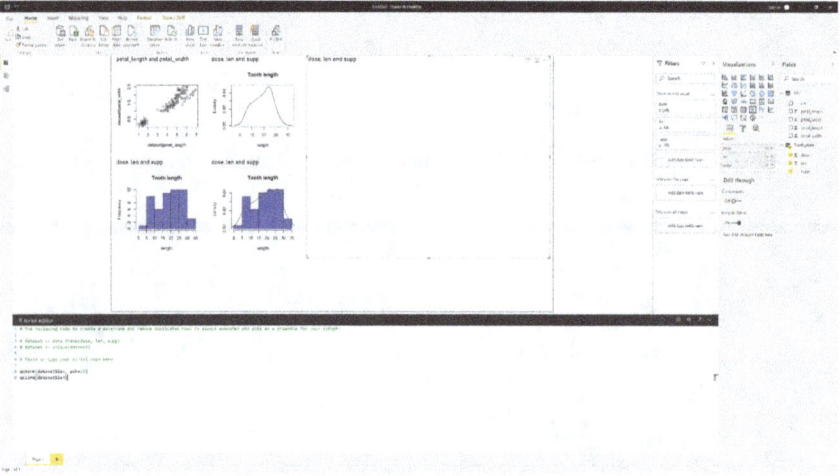

**Figure 12-18**   ToothGrowth Q-Q Plot

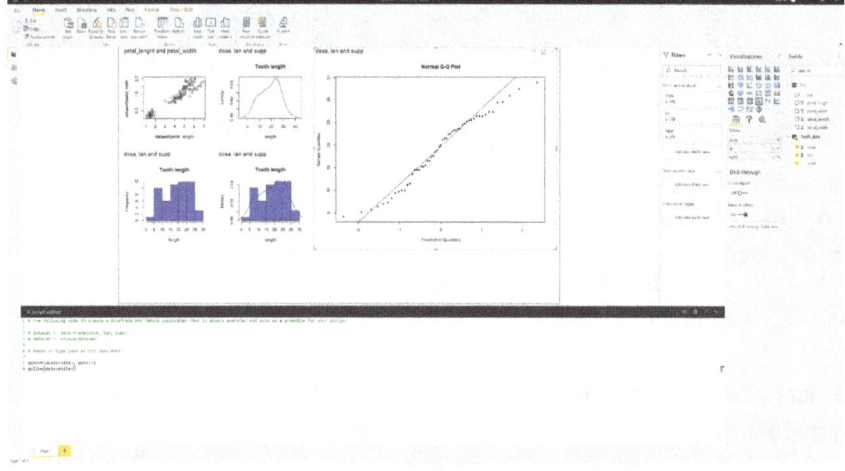

**Figure 12-19**   ToothGrowth Q-Q plot

**Source (12-19)**

```
# dataset <- data.frame(dose, len, supp)
# dataset <- unique(dataset)

qqnorm(dataset$len, pch = 20)
qqline(dataset$len)
```

## 12.3.2 *Categorical Univariate Plot*

Find additional support for inserting data by clicking on the "Get Data" menu, as shown in Figure 12-20.

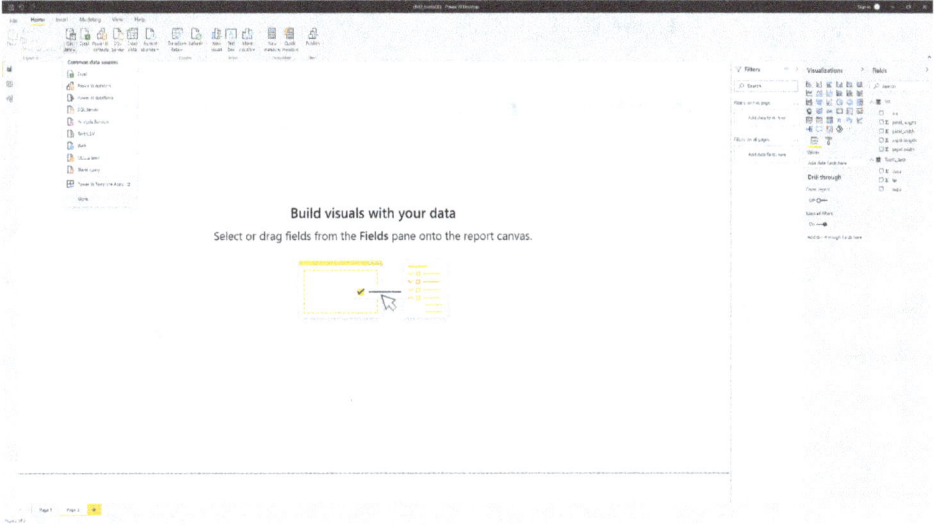

**Figure 12-20**   Additional data import method

Search and click on the "R Script" icon as shown in Figure 12-21.

Generate sample data to create a categorical univariate plot. Enter the following source in the R Script window, as shown in Figure 12-22.

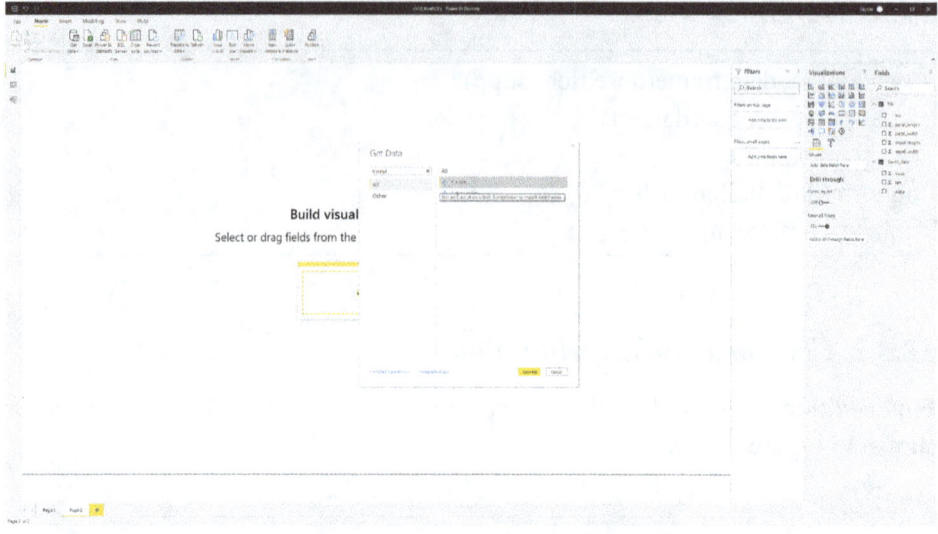

**Figure 12-21**  R Script data import method

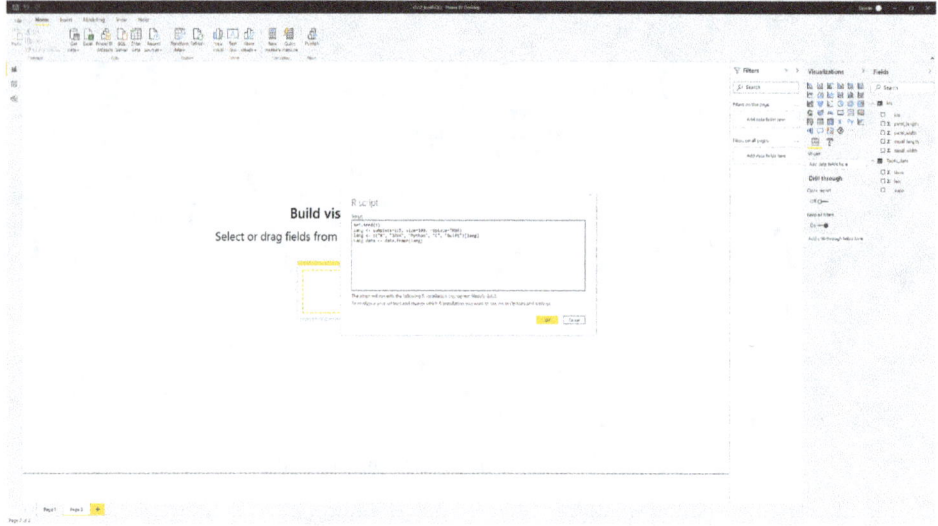

**Figure 12-22**  Generate sample data

**Source (12-22)**

```
set.seed(1)
lang <- sample(x=1:5, size=100, replace=TRUE)
lang <- c("R", "JAVA", "Python", "C", "Swift")[lang]
Lang_data <- data.frame(lang)
```

After loading, you can check the created sample data, as shown in Figure 12-23.

After entering the source, you can see that the graph is not plotted normally, as shown in Figure 12-24.

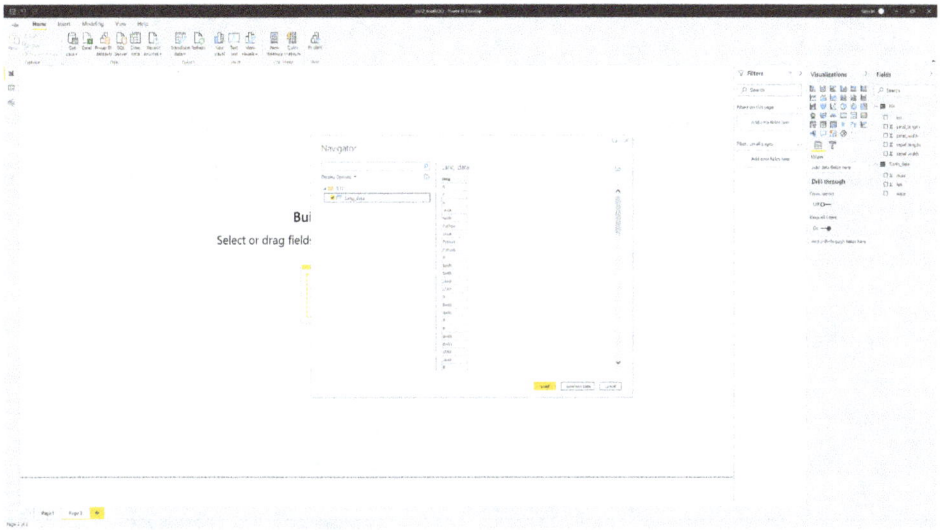

**Figure 12-23**  Sample data generation complete

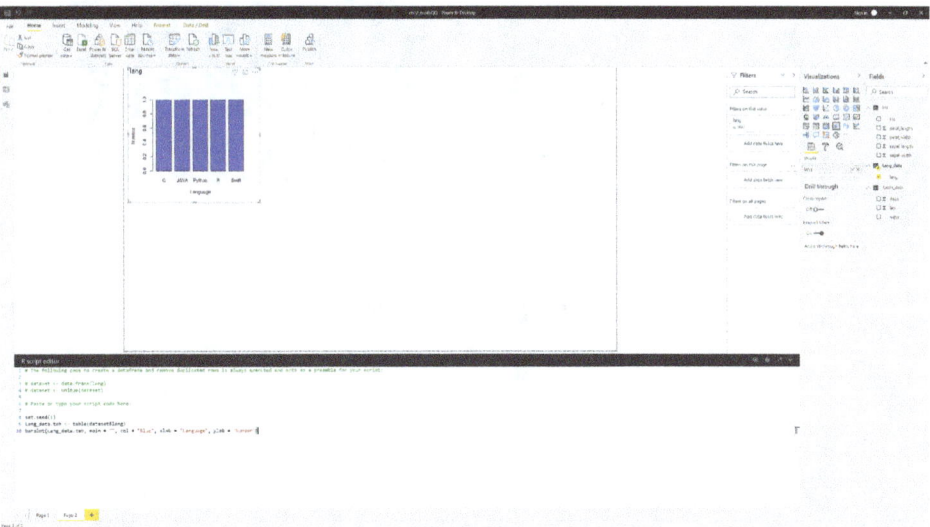

**Figure 12-24**  R Script error in Power BI

**Source (12-24)**

```
# dataset <- data.frame(lang)
# dataset <- unique(dataset)

set.seed(1)
Lang_data.tab <- table(Lang_data)
barplot(Lang_data.tab,main="",col="Blue",xlab="Language",ylab="Number")
```

However, in Figure 12-25, you can see that R Studio is functioning normally.

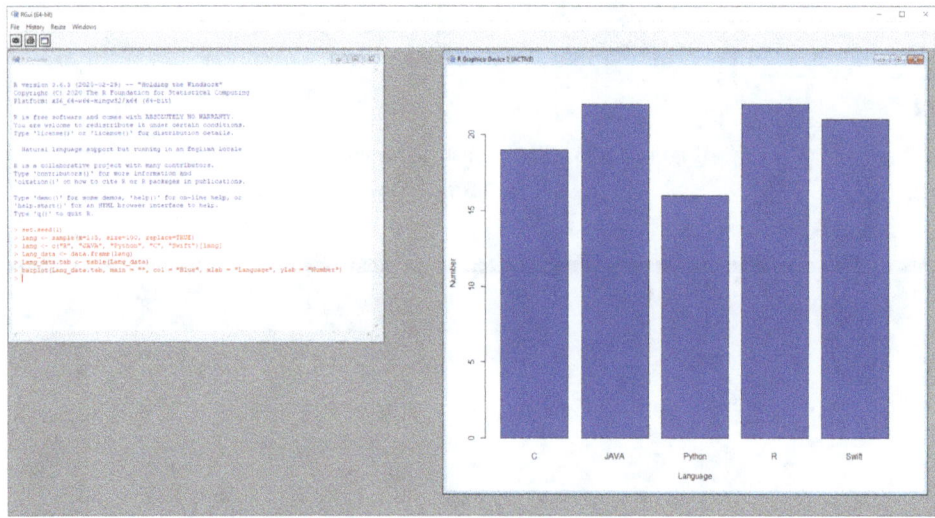

**Figure 12-25**   R Studio

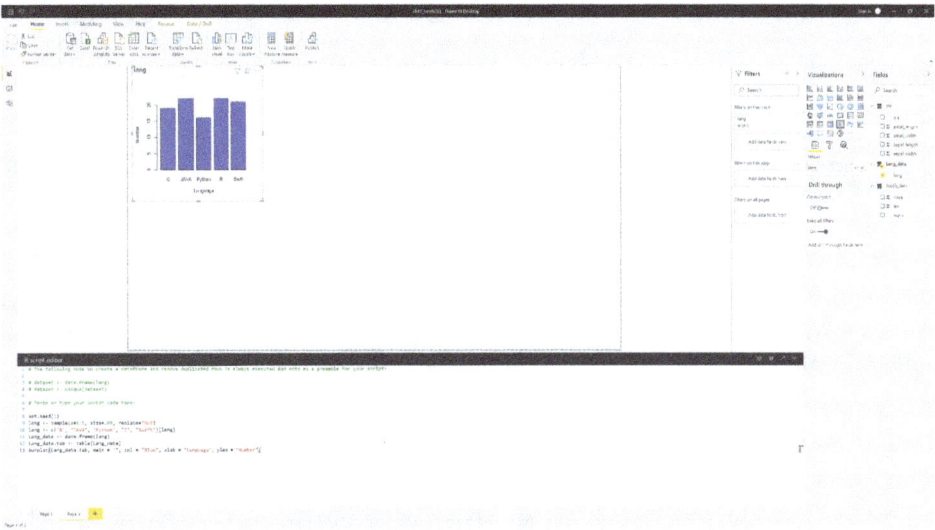

**Figure 12-26** Language data column chart

**Source (12-26)**

```
# dataset <- data.frame(lang)
# dataset <- unique(dataset)

set.seed(1)
lang <- sample(x=1:5, size=100, replace=TRUE)
lang <- c("R", "JAVA", "Python", "C", "Swift")[lang]
Lang_data <- data.frame(lang)
Lang_data.tab <- table(Lang_data)
barplot(Lang_data.tab,main="",col="Blue",xlab="Language",ylab="Number")
```

Enter the following Source in the R Script window, as shown in Figure 12-26. Because of errors caused by changes in the structure of the data in the import process, let's utilize the source of R Studio as a means of solving this problem

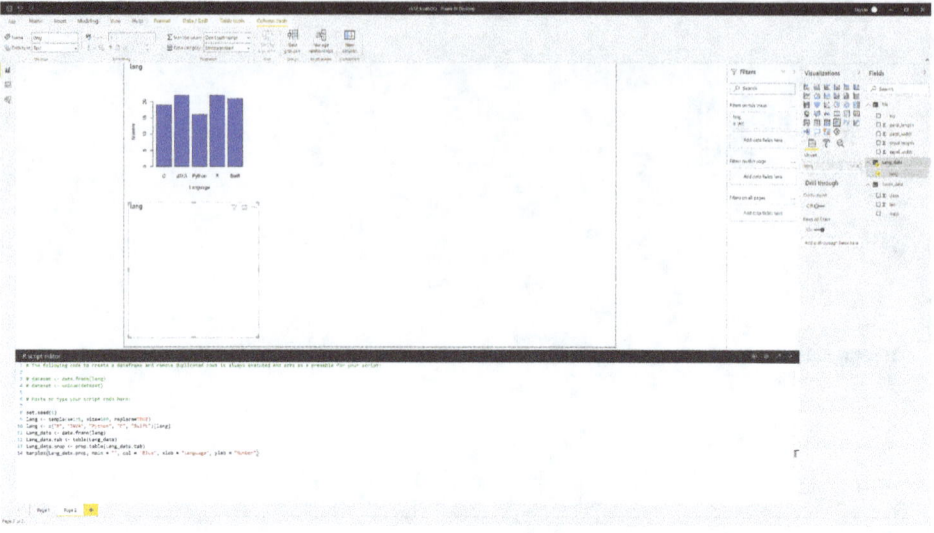

**Figure 12-27**   Language data relative ratio column chart

You can check that the column chart is plotting normally. Enter the source below in the R Script window, as shown in Figure 12-27 to draw the relative ratio column chart.

After executing the R Script, you can see that the relative ratio column chart has been created, as shown in Figure 12-28.

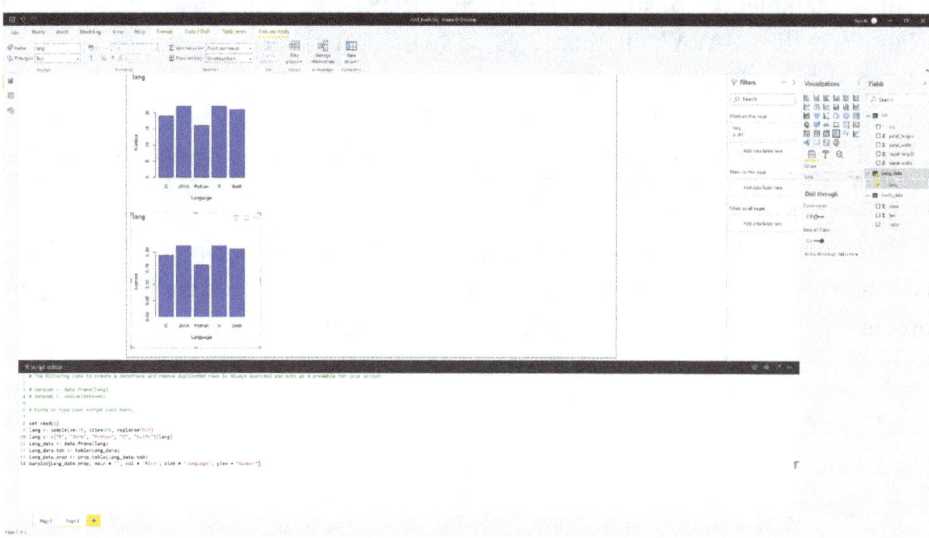

**Figure 12-28**   Language data relative ratio column chart

**Source (12-28)**

```
# dataset <- data.frame(lang)
# dataset <- unique(dataset)

set.seed(1)
lang <- sample(x=1:5, size=100, replace=TRUE)
lang <- c("R", "JAVA", "Python", "C", "Swift")[lang]
Lang_data <- data.frame(lang)
Lang_data.tab <- table(Lang_data)
Lang_data.prop <- prop.table(Lang_data.tab)
barplot(Lang_data.prop, main="",col="Blue",xlab="Language",
ylab="Number")
```

Pie charts are useful in determining the distribution of categorical data. Enter the following source in the R Script window, as shown in Figure 12-29.

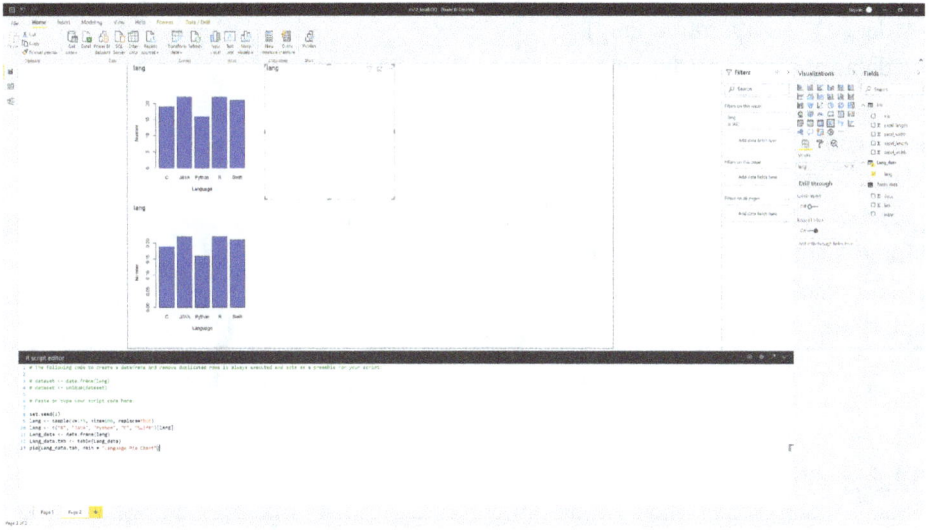

**Figure 12-29** Language data pie chart

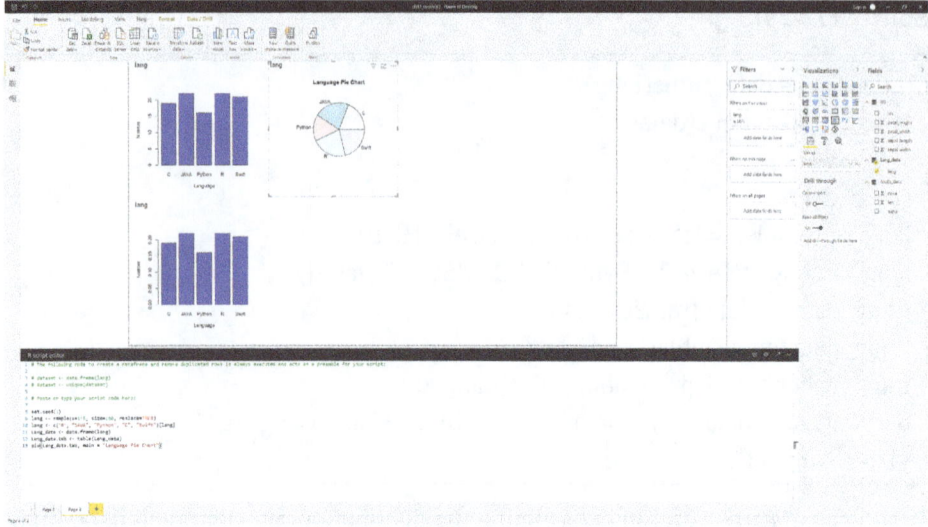

**Figure 12-30**  Language data pie chart

**Source (12-30)**

```
# dataset <- data.frame(lang)
# dataset <- unique(dataset)

set.seed(1)
lang <- sample(x=1:5, size=100, replace=TRUE)
lang <- c("R", "JAVA", "Python", "C", "Swift")[lang]
Lang_data <- data.frame(lang)
Lang_data.tab <- table(Lang_data)
pie(Lang_data.tab, main="Language Pie Chart")
```

You can see that the pie chart has been created, as shown in Figure 12-30.

## 12.3.3 *Numerical Bivariate Plot*

Before drawing a numerical bivariate plot, you need to install a package to import the Galton data. Enter the following code in R Studio, as shown in Figure 12-31. Set the appropriate region for downloading, and the download will begin.

After the download is completed, enter the following source into R Studio, as shown in Figure 12-32. You can check the structure of the data through the dim () function.

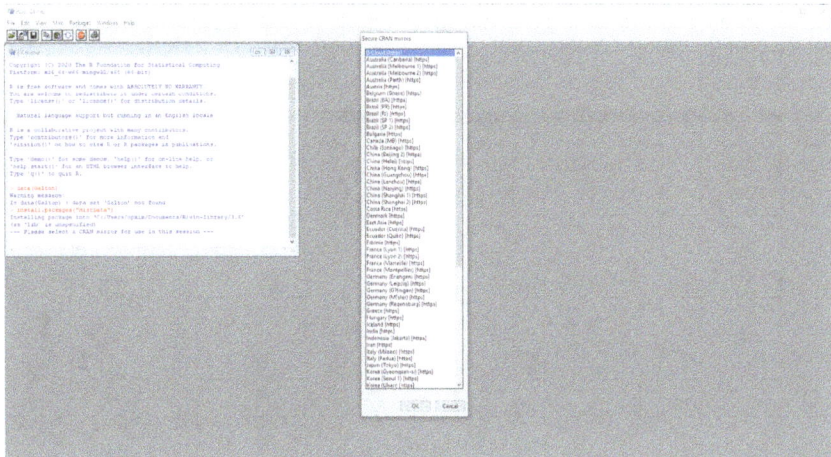

**Figure 12-31**  Package installation for inserting Galton data

## Source (12-31)

```
install.packages("HistData")
```

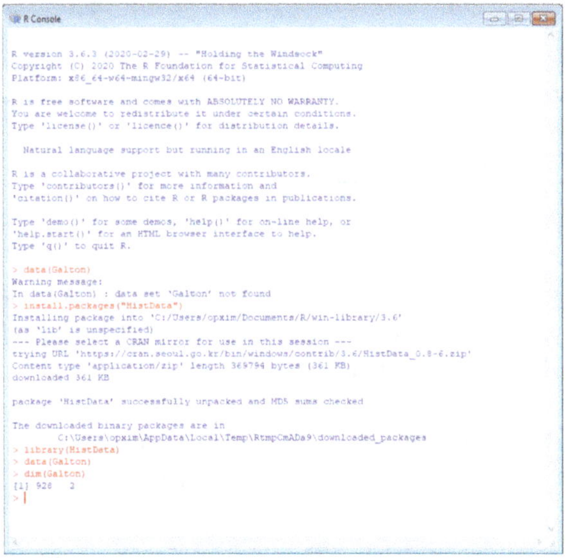

**Figure 12-32**  Package installation for inserting Galton data

## Source (12-32)

```
library (HistData)
data (Galton)
dim (Galton)
```

Import the Galton data by entering the following source in the R Script, as shown in Figure 12-33.

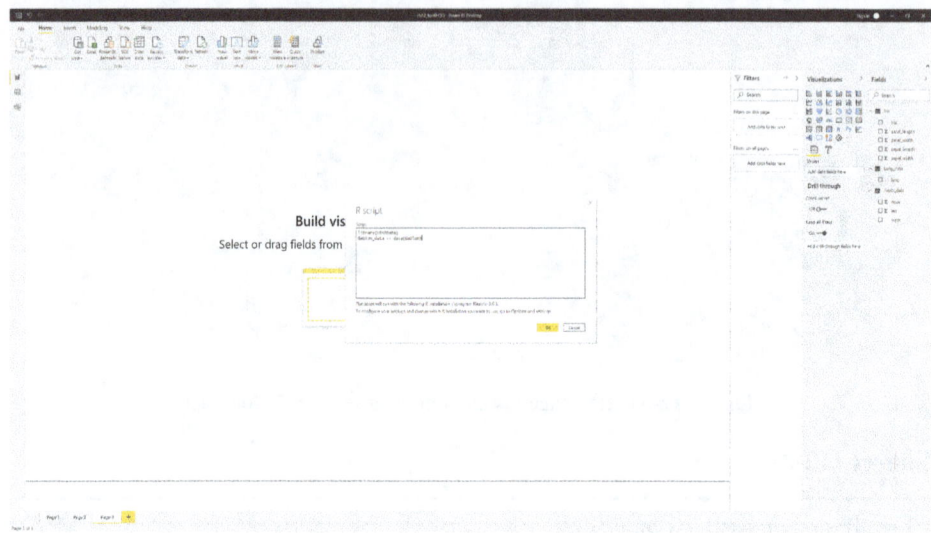

**Figure 12-33**    Import Galton data with R Script

**Source (12-33)**

Library (HistData)
Galton_data <- data (Galton)

- Galton data

Galton (1886) presented these data in a table, showing a cross-tabulation of 928 adult children born to 205 fathers and mothers, by their height and their mid-parental height. He visually smoothed the bivariate frequency distribution and showed that the contours formed concentric and similar ellipses, thus setting the stage for correlation, regression and the bivariate normal distribution. A data frame was created with 928 observations on the following 2 variables:

1. parent a numeric vector: height of the mid-parent (average of father and mother)
2. child a numeric vector: height of the child

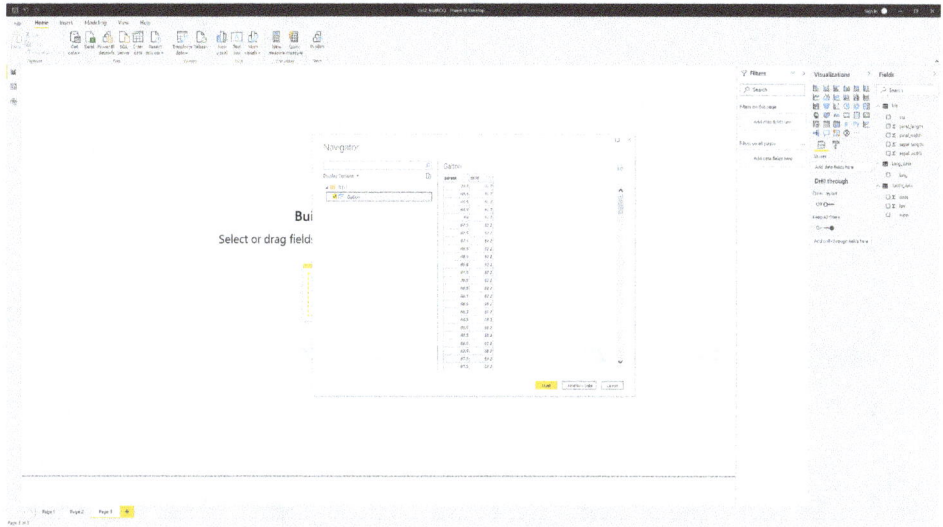

**Figure 12-34** Galton data importing success

After loading, you can check the Galton data, as shown in Figure 12-34.

Enter the source below in the R Script, as shown in Figure 12-35, to analyze the relationship between the parent's and the child's heights.

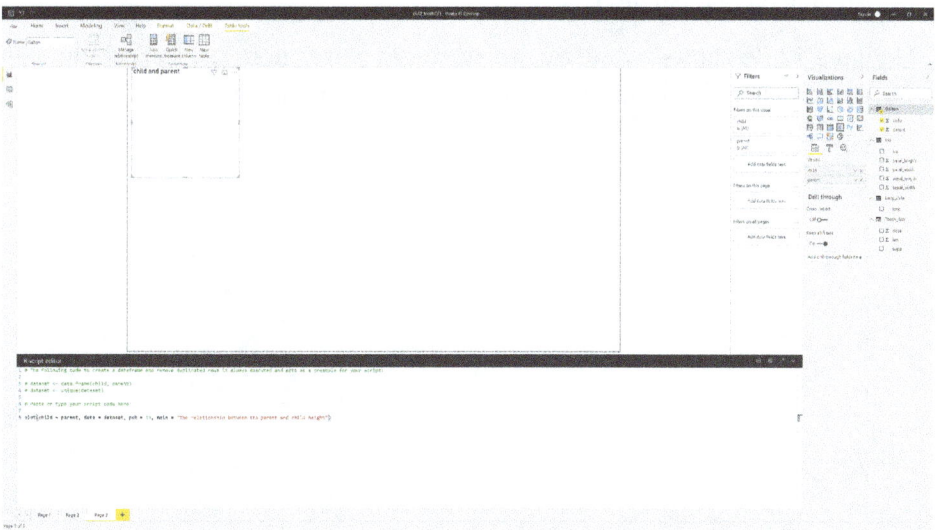

**Figure 12-35** The relationship of height between a parent and a child

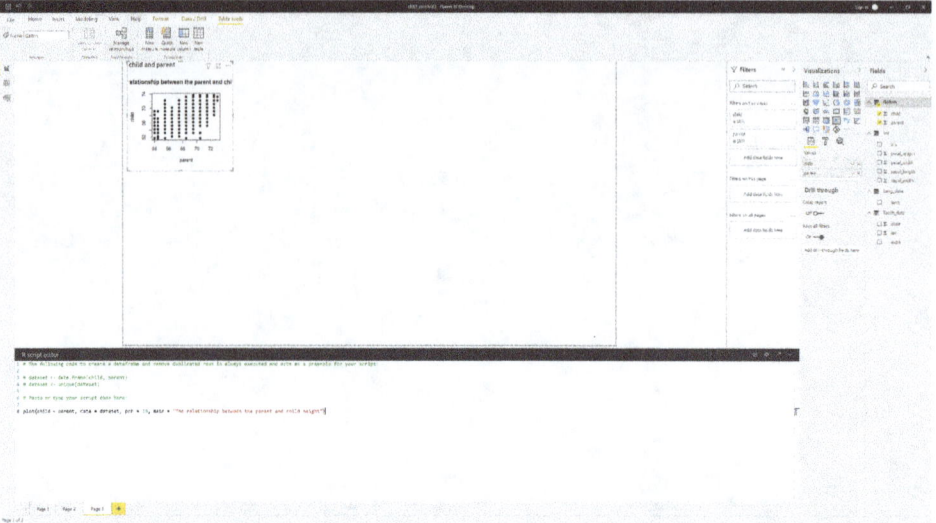

**Figure 12-36**   The relationship of height between a parent and a child

## Source (12-36)

```
# dataset <- data.frame(child, parent)
# dataset <- unique(dataset)

plot(child ~ parent, data = dataset, pch = 19, main = "The relationship between
the parent and child height")
```

Figure 12-36 shows indirectly that there is a correlation. Let's check through a linear regression line to make sure that there is a correlation.

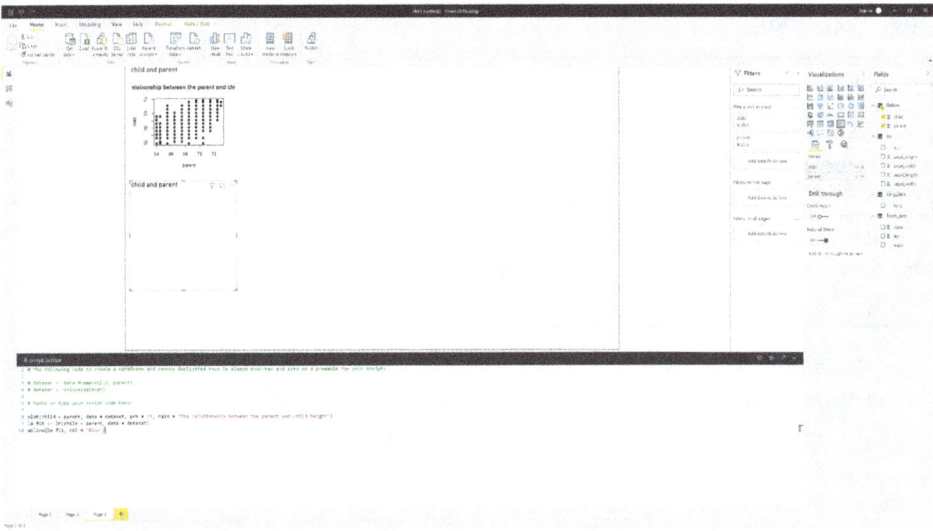

**Figure 12-37**   The regression line of height between a parent and a child

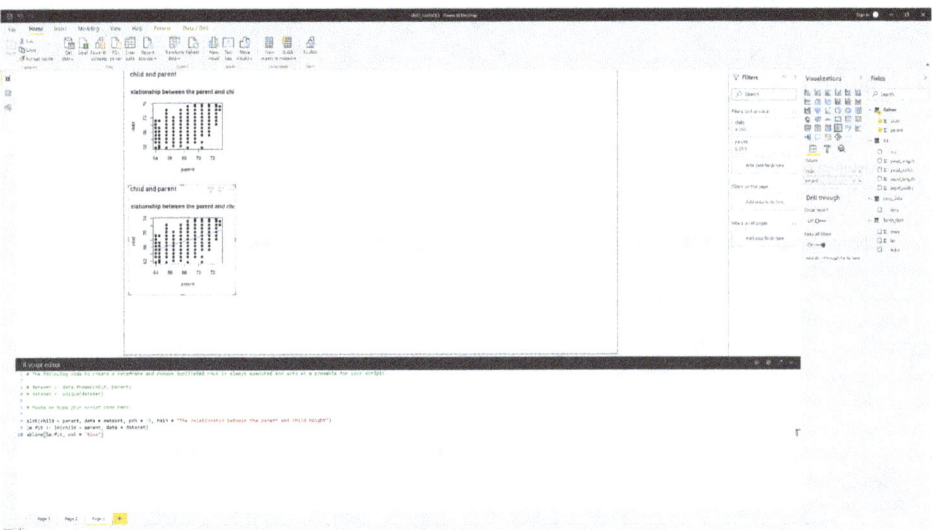

**Figure 12-38**   The regression line of height between a parent and a child

Enter the source below in the R Script window, as shown in Figure 12-37, to draw the linear regression line.

The result in Figure 12-38 confirms that there is a positive correlation.

**Source (12-38)**

```
# dataset <- data.frame(child, parent)
# dataset <- unique(dataset)

plot(child ~ parent, data = dataset, pch = 19, main = "The relationship between
the parent and child height")
lm.fit <- lm(child ~ parent, data = dataset)
abline(lm.fit, col = "Blue")
```

　　The previous figures do not seem to represent all the 928 data points. Since this is superimposed in the same area, let's scatter the scatter plot. Enter the following source in the R Script window, as shown in Figure 12-39.

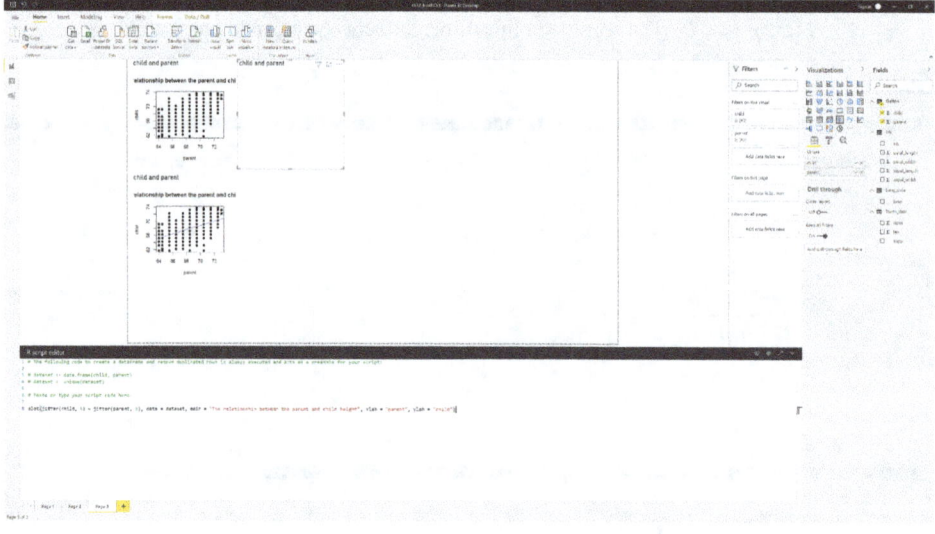

**Figure 12-39**　A scatterplot of height between parents and children

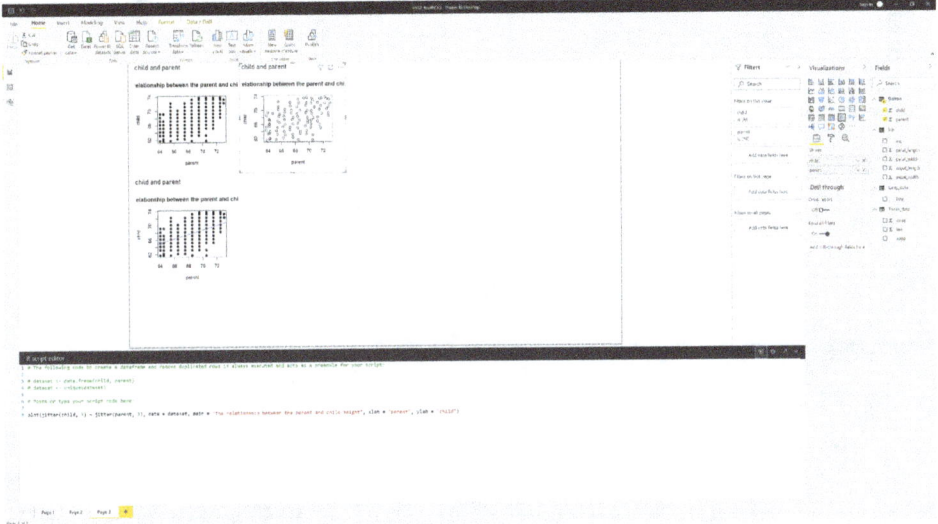

**Figure 12-40**   A scatter plot of height between parents and children

## Source (12-40)

```
# dataset <- data.frame(parent, child)
# dataset <- unique(dataset)

plot(jitter(child, 3) ~ jitter(parent, 3), data = dataset, main = "The relationship
between the parent and child height", xlab="parent", ylab="child")
```

Figure 12-40 shows scattered points on the scatter plot.

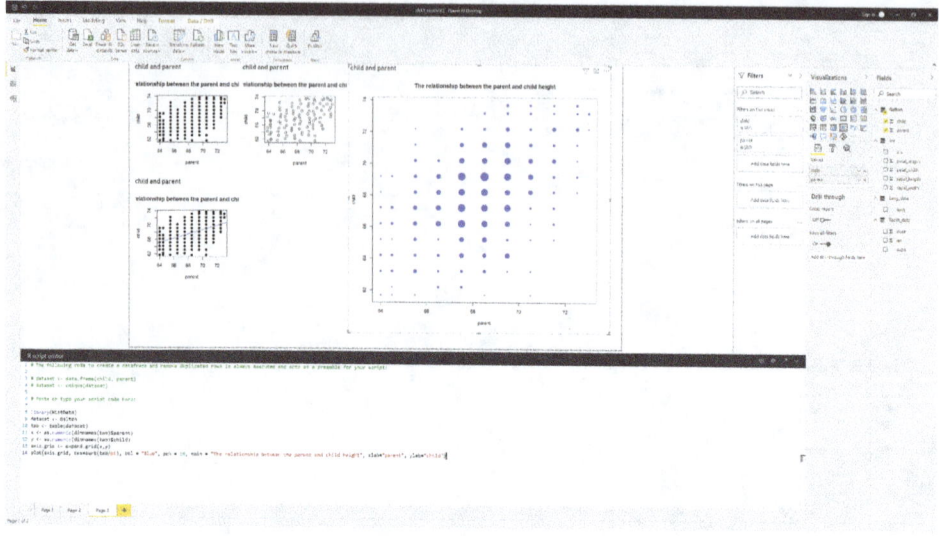

**Figure 12-41**   Scatter plot by point size

**Source (12-41)**

```
# dataset <- data.frame(child, parent)
# dataset <- unique(dataset)

library(HistData)
dataset <- Galton
tab <- table(dataset)
x <- as.numeric(dimnames(tab)$parent)
y <- as.numeric(dimnames(tab)$child)
axis.grid <- expand.grid(x,y)
plot(axis.grid, cex=sqrt(tab/pi), col = "Blue", pch = 19, main = "The relationship
between the parent and child height", xlab="parent", ylab="child")
```

Figure 12-41 shows the size of the points in proportion to their frequency. The chart is represented in the same way as in Figure 12-41. It is called a bubble chart. To apply a bubble chart, enter the following source in the R Script window, as shown in Figure 12-41.

## 12.3.4 *Categorical Bivariate Plot*

Before drawing a categorical bivariate plot, let's take a look at the structure by importing the Titanic data provided by R Studio.

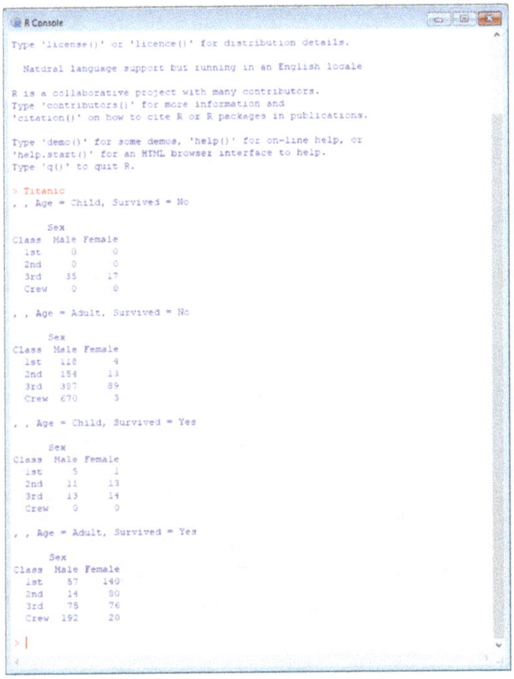

**Figure 12-42** Titanic data structure

## Source (12-42)

Titanic

Enter the source below in the R Studio, as shown in Figure 12-42.

• Titanic data

The Titanic data set provides information on the fate of passengers on the fatal maiden voyage of the ocean liner "Titanic", summarized according to economic status (class), sex, age and survival.

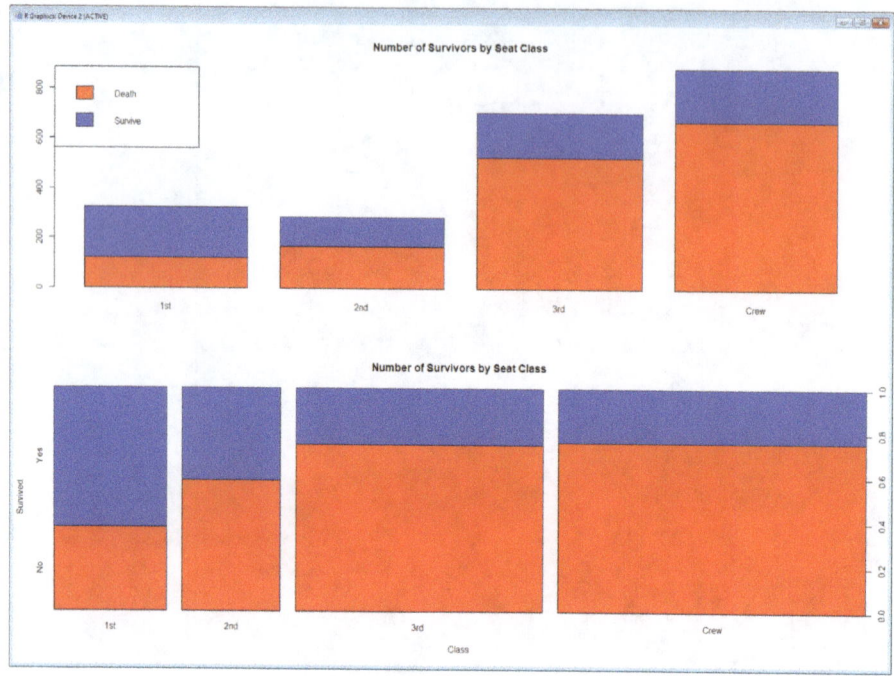

**Figure 12-43**    Titanic data stacked column chart

## Source (12-43)

```
par(mfrow = c(2,1))
barplot(apply(Titanic, c(4,1), sum), col = c("Red", "Blue"), main = "Number of
Survivors by Seat Class")
legend("topleft", fill = c("Red", "Blue"), c("Death", "Survive"))
spineplot(margin.table(Titanic, c(1,4)), col = c("Red", "Blue"), main = "Number
of Survivors by Seat Class")
```

Figure 12-43 shows the number of survivors and fatalities according to class, as well as the ratio. Enter the source below into R Studio to implement.

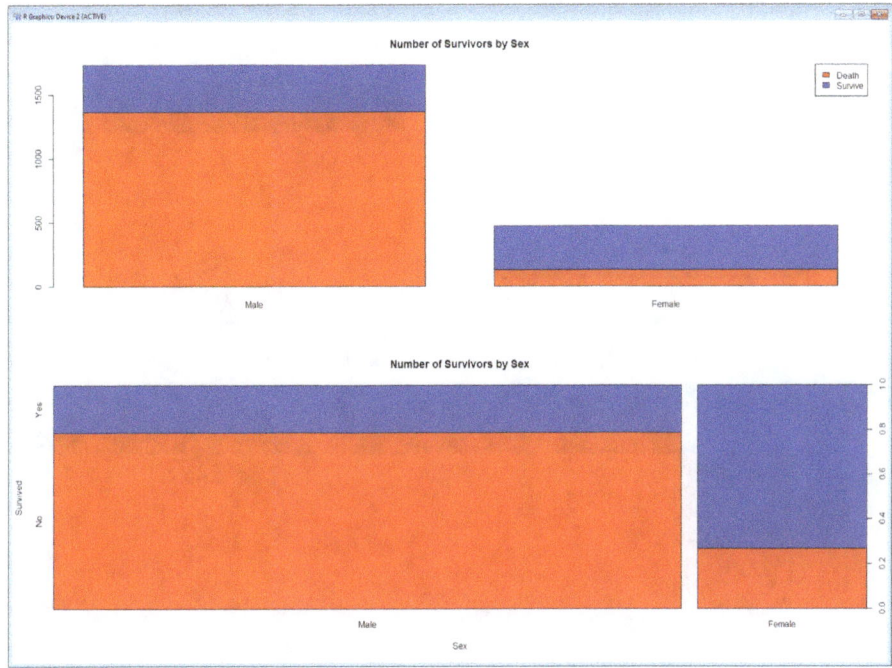

**Figure 12-44**   Titanic data stacked column chart

## Source (12-44)

```
par(mfrow = c(2,1))
barplot(apply(Titanic, c(4,2), sum), col = c("Red", "Blue"), main = "Number of
Survivors by Sex")
legend("topright", fill = c("Red", "Blue"), c("Death", "Survive"))
spineplot(margin.table(Titanic, c(2,4)), col = c("Red", "Blue"), main = "Number
of Survivors by Sex")
```

Figure 12-44 shows the number of survivors and fatalities according to sex, as well as the ratio. Enter the source below into R Studio to implement.

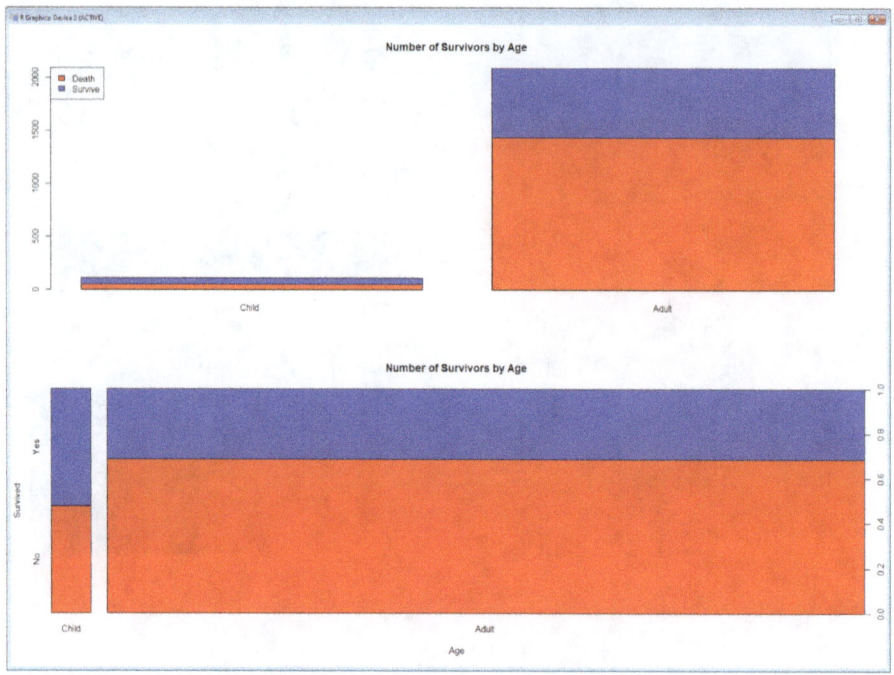

**Figure 12-45**   Titanic data stacked column chart

## Source (12-45)

```
par(mfrow = c(2,1))
barplot(apply(Titanic, c(4,3), sum), col = c("Red", "Blue"), main = "Number of
Survivors by Age")
legend("topleft", fill = c("Red", "Blue"), c("Death", "Survive"))
spineplot(margin.table(Titanic, c(3,4)), col = c("Red", "Blue"), main = "Number
of Survivors by Age")
```

Figure 12-45 shows the number of survivors and fatalities according to age, as well as the ratio. Enter the source below into R Studio to implement.

## 12.4  Practice Questions

Q1. Explain the difference between numerical data and categorical data.



Q2. Explain the difference between univariate data and bivariate data.



Q3. Explain the Q-Q plot (i.e. the Quantile-Quantile plot).

# Bibliography

ACM SIGKDD. *"Data Mining Curriculum: A Proposal" Last modified April 30, 2006.* Accessed January 27, 2014. *https://www.kdd.org/curriculum/view/introduction*

Altman, Naomi S. *"An introduction to kernel and nearest-neighbor nonparametric regression." The American Statistician, 46 (3) (1992): 175–185. doi:10.1080/00031305.1992.10475879.*

"Artificial intelligence: Google's AlphaGo beats Go master Lee Se-dol". BBC News. Last modified March 12, 2016. Accessed March 17 2016.

Ben-Hur, Asa, David Horn, Hava Siegelman and Vladimir N. Vapnik. "Support vector clustering." *Journal of Machine Learning Research, 2 (2001): 125–137.*

Cho, Young Im. *Artificial Intelligence System.* Hong Reung Publishing Company, 2012. Korea

Cortes, Corinna and Vladimir N. Vapnik. "Support-vector networks." *Machine Learning, 20(3) (1995): 273–297.* CiteSeerX 10.1.1.15.9362. doi:10.1007/BF00994018.

DeepMind, https://deepmind.com/ Accessed June 29, 2020.

Goki, Saito. *Deep Learning from the Bottom.* Hanbit Media, 2017. Korea

"Google I/O 2016 — Keynote", https://www.youtube.com/watch?v=862r3XS2YB0&t=7190 Accessed June 29, 2020.

"Google achieves AI 'breakthrough' by beating Go champion". BBC News. Last modified January 27, 2016.

Hale, Mike. "Actors and Their Roles for $300, HAL? HAL!". *New York Times,* February 8, 2011.

*IBM Corporation, "DeepQA Project: FAQ",* https://researcher.watson.ibm.com/researcher/view_group.php?id=2099 accessed June 29, 2020.

JDM. "JDM's Blog" Last modified April 9, 2015. http://jdm.kr/blog/112

Kim, Euijoong. *Introduction to AI, Machine Learning, and Deep Learning with Algorithms.* Wikibooks, 2017. Korea

Kim, Seong-pil. *Deep Learning First Step.* Hanbit Media, 2016. Korea

Last modified April 1, 2017, Byung ho Kang, https://www.slideshare.net/medit74/ss-74123546?qid=5a76f350-f606-4cb0-aa56-7527fc1d7a67&v=&b=&from_search=2

Last modified August 3, 2016. NVDIA KOREA, http://blogs.nvidia.co.kr/2016/08/03/difference_ai_learning_machinelearning/

Last modified February 9, 2017. Nomorebets, http://blog.naver.com/PostView.nhn?blogId=nomore_bet&logNo=220930957727

Last modified January 27, 2016. Google Research Blog, "AlphaGo: Mastering the ancient game of Go with Machine Learning" https://ai.googleblog.com/2016/01/alphago-mastering-ancient-game-of-go.html

Last modified January 28, 2016., Chris Duckett, "Google AlphaGo AI clean sweeps European Go champion", ZDNet

Last modified June 10, 2020. Wikipedia. "Monte Carlo tree search." https://en.wikipedia.org/wiki/Monte_Carlo_tree_search

Last modified June 14, 2017. Kim, Peter, https://projectresearch.co.kr/2017/06/14/%EB%A8%B8%EC%8B%A0%EB%9F%AC%EB%8B%9Dml%EC%9D%98-%EA%B0%84%EB%9E%B5%ED%95%9C-%EC%97%AD%EC%82%AC/

Last modified June 3, 2020. Wikipedia. "Watson (computer)." https://en.wikipedia.org/wiki/Watson_(computer)

Last modified June 9, 2017. pubdata, http://pubdata.tistory.com/134

Last modified March 11, 2016. Beomsu Kim, https://shuuki4.wordpress.com/2016/03/11/alphago-alphago-pipeline-%ed%97%a4%ec%a7%91%ea%b8%b0/

Last modified May 3, 2016., Sarah Griffiths "Artificial intelligence breakthrough as Google's software beats grandmaster of Go, the 'most complex game ever devised'", dailymail

Last modified May 6, 2016. SK C&C, https://blog.skcc.com/2808

Last modified May 7, 2020. Microsoft, https://docs.microsoft.com/ko-kr/azure/machine-learning/studio/algorithm-choice

Last modified June 21, 2020. Wikipedia. "Minimax." https://en.wikipedia.org/wiki/Minimax

Last modified June 27, 2020. namuwiki, https://namu.wiki/w/%EC%95%8C%ED%8C%8C%EA%B3%A0

Marsland, Stephen. *Machine Learning: An Algorithm Perspective, Second Edition*. CRC Press, 2014. USA

"Match 1 — Google DeepMind Challenge Match: Lee Sedol vs AlphaGo". https://www.youtube.com/watch?v=vFr3K2DORc8 Accessed June 29, 2020.

*McCarthy, John, Marvin Minsky Nathan Rochester, and Claude Shannon (1955),* "A Proposal for the Dartmouth Summer Research Project on Artificial Intelligence.", Accessed June 29, 2020. http://www-formal.stanford.edu/jmc/history/dartmouth/dartmouth.html

McCorduck, Pamela. *Machines Who Think, Second Edition*. A.K. Peters, 2004. USA

Metz, Cade. "In Major AI Breakthrough, Google System Secretly Beats Top Player at the Ancient Game of Go". *WIRED, January 27, 2016.* Accessed February 1, 2016.

Michalski, Ryszard S., Jaime G. Carbonell, and Tom M. Mitchell (eds.). *Machine Learning: An Artificial Intelligence Approach.* Springer Science & Business Media, 2013. Germany

Nillson, Nils J. *Artificial Intelligence: A New Synthesis.* Elsevier, 1998. Netherlands

Russell, Stuart, and Peter Norvig. "Chapter 1.1: What is AI?" in *Artificial Intelligence: A Modern Approach*, 1–4. Malaysia: Pearson Education, 2016. https://people.eecs. berkeley.edu/~russell/intro.html

Shai, Shalev-Shwartz and S. Ben-David Shai. *Understanding Machine Learning: From Theory to Algorithms.* Cambridge: Cambridge University Press, 2014. UK

Silver, David, Aja Huang, Chris J. Maddison, Arthur Guez, Laurent Sifre, George van den Driessche, Julian Schrittwieser, Ioannis Antonoglou, and Veda Panneershelvam. "Mastering the game of Go with deep neural networks and tree search." *Nature, 529 (7587) (2016): 484–489.* doi:10.1038/nature16961.

Su, Steph. "The Jeopardy! IBM Challenge and What It Tells Us About Language" Last modified February 25, 2011. http://stephsureads.blogspot.com/2011/02/jeopardy-ibm-challenge-and-what-it.html

Yang, Ki-cheol. *Artificial Intelligence Theory and Practice.* Hong Reung Publishing Company, 2018. Korea